The Dynamics of Sustainable Innovation Journeys

This book shows that sustainable development should be analysed and managed as an innovation journey in which social, technological, political and cultural dimensions become aligned. The 'journey' aspect captures the open and uncertain nature of sustainable developments and highlights the agency dimension, with actors navigating, negotiating, groping and struggling their way forward (and sometimes backward). The book addresses the following research questions: What are the key processes and micro-dynamics of innovation journeys? Which policy lessons can be drawn for managing sustainable innovation journeys? To conceptualize the multi-dimensional nature of innovation journeys the book draws on insights from industrial economics, evolutionary economics, sociology of technology, political science and cultural studies. The book develops several new conceptual frameworks that make different crossovers between these disciplines. These frameworks are empirically tested with case studies on biofuels, onshore wind power, low energy housing, photovoltaic solar cells, biomass and fuel cells. The empirical studies are also used to derive several robust lessons as to how policy makers can influence sustainable innovation journeys.

This book was published as a special issue of *Technology Analysis & Strategic Management*.

Frank W. Geels is Professor at SPRU, one of the founding institutes of innovation studies, at the University of Sussex. He is a world-leading scholar on socio technical transitions to sustainability, and is well known for his work on the multi-level perspective and empirical studies of a dozen historical and future transitions.

Marko P. Hekkert is full professor at Innovation Studies, Copernicus Institute of Sustainable Development, Utrecht University in The Netherlands. He is research director of the Copernicus research institute and head of innovation studies where 35 scholars focus on understanding emerging technologies, innovation system dynamics and sustainability transitions.

Staffan Jacobsson is Professor in Science and Technology Policy at Chalmers University of Technology, Gothenburg, Sweden. His research interest lies in (i) Science Policy, (ii) Energy Policy and (iii) how we may conceptualize and empirically study the dynamics of technological innovation systems. He has been honored with three teaching awards.

The Dynamics of Sustainable Innovation Journeys

Edited by
Frank W. Geels, Marko P. Hekkert and Staffan Jacobsson

Routledge
Taylor & Francis Group
LONDON AND NEW YORK

First published 2011
by Routledge
4 Park Square, Milton Park, Abingdon, Oxon, OX14 4RN

Simultaneously published in the USA and Canada
by Routledge
605 Third Avenue, New York, NY 10017

Routledge is an imprint of the Taylor & Francis Group, an informa business

This book is a reproduction of *Technology Analysis & strategic Management* 20.5. The Publisher requests that those citing this book use the bibliographical details of the journal issue on which the book is based.

Typeset in Times New Roman by Taylor & Francis Books.

British Library Cataloguing in Publication Data
A catalogue record for this book is available from the British Library

ISBN 13: 978-0-415-61866-3 (hbk)

Disclaimer
The publisher would like to make readers aware that the chapters in this book are referred to as articles as they had been in the special issue. The publisher accepts responsibility for any inconsistencies that may have arisen in the course of preparing this volume for print.

Contents

The dynamics of sustainable innovation journeys

1. The contribution of innovation studies to the solution of persistent environmental problems

Since the 1970s, environmental problems have risen on the political agenda. Many environmental problems (such as water pollution, local air pollution, acid rain) have since been solved or substantially reduced by policies and incremental clean technologies (e.g. fuel gas desulphurisation in power plants, three-way catalytic converters in cars), but other problems have grown worse, e.g. climate change, loss of biodiversity, deforestation, depletion of fish stocks and of clean water supplies. The incremental approach is less likely to work with these problems. With regard to climate change, for instance, global CO_2 emissions must be cut by 80% by 2050 if the risks are to remain within acceptable limits (Stern 2006). It is difficult to achieve such cuts with incremental technical changes. Hence, more radical 'green' innovations are needed.

Although 'green' technologies with less CO_2 emissions have emerged, their further development and the transformation of (fossil-fuel based) energy system into more sustainable directions will require a massive effort. This effort is massive for three reasons. First, the global energy system is extremely large and growing rapidly. The global power production increased from 6000 TWh in 1973 to 17,000 TWh in 2004, and is expected to grow to 25,000–35,000 TWh in 2050. Even with fossil-fuel or nuclear technologies it will be difficult to meet this growing demand. To transform this expanding energy system into carbon neutral directions will require even more effort, including large investments in the creation of new industrial structures. Second, the time scale for these changes is relatively short. Historical studies show that large scale transformation processes often takes many decades to unfold (Carlsson and Jacobsson 1997; Geels 2002). The challenge is to stabilise atmospheric CO_2 concentrations in the next four decades (by 2050). This is difficult to achieve, because of timelags in the climate system. Analysts therefore estimate that we have a change window of about a decade to start reducing CO_2 emissions (Stern 2006). Giving the previously mentioned expansion of energy demand, this time period forms an unprecedented challenge. Third, 'green' innovations face several obstacles that hinder rapid diffusion. Schumpeter (1934, 86–87) already noted that most innovations with a substantial degree of novelty face such problems:

> In the breast of one who wishes to do something new, the forces of habit rise up and bear witness against the embryonic project .... The reaction of the social environment against one who wishes to do something new, may manifest itself first of all in the existence of legal or political impediments. .... Any deviating conduct by a member of a social group is condemned . ... Even a mere astonishment at the deviation ... exercises a pressure on the individual.

Such obstacles are particularly large in the case of the energy system. Unruh (2000) therefore appropriately characterised the current situation as a 'carbon lock-in'. We mention three prominent problems.

(1) Initially, sustainable technologies tend to be more expensive and have lower performance (in mainstream dimensions) than existing technologies. Usually, market niches provide early footholds for radical innovations, with particular users accepting teething problems because the innovations offer advantages in that application domain. This mechanism is more complicated for 'green' innovations, because a clean environment is a collective good. 'Green' innovations provide benefits (reduced emissions) for society *at large* while costs are borne by individual users. The resulting (free rider) problems to form early niche markets hinder the innovation process (Jacobsson and Bergek 2004).
(2) Uncertainties about future market and regulations also hinder the commitment of firms to the development of sustainable technologies. On the one hand, many large energy firms and car manufacturers *do* invest in 'green' technologies, because they recognise the climate change problem (e.g. Shell in hydrogen, Toyota in hybrid cars, Ford in fuel cell cars). On the other hand, they do not (yet) fully commit to these innovations, because of market uncertainties and fear of cannibalising their existing products. Many green innovations therefore remain on the shelf or are paraded only on demonstration shows. The continuation of uncertainties prevents firms from strong commitment to the development and marketing of 'green' innovations.
(3) Existing technologies and sociotechnical systems are stabilised by lock-in mechanisms (Walker 2000; Unruh 2000). Long periods of dynamic increasing return (e.g. learning by doing and using, scale economies, network externalities) put them in advantageous positions. Standards, favourable regulations, sunk investments (in capital, competencies, social networks, infrastructures) and vested interests also provide existing technologies with stability. Subsidies (e.g. for coal), user lifestyles and behavioural patterns may provide further stability. Sustainable technologies may, therefore, face additional barriers when they have a 'mis-match' with aspects of existing systems (Freeman and Perez 1988). In sum, 'green' innovations do not compete with existing technologies on 'level playing fields'.

Continuing on the above discussion, we conclude that three often mentioned response strategies to environmental problems are insufficient. We briefly discuss these strategies and then introduce the fourth approach, which is central to this special issue.

(1) *Neo-liberal strategies* focus on 'getting the prices right'. Neo-liberal thinkers argue that environmental problems lead to scarcity, which translates into higher prices, which will trigger changes in the behaviour of consumers (demand for 'sustainable' products) and firms (investments to develop these products). Market failures may occur for collective goods, which require governments to introduce measures that internalise external costs (e.g. taxes, tradeable

emission permits). For some environmental problems and under certain conditions (rational agents, full information, perfect markets) this approach can be effective and efficient, but it is more problematic for radical innovations and transitions, which are characterised by uncertainty about technologies, user preferences, and market institutions. In such conditions, rational calculations are likely to be less prominent than search processes, power struggles, learning and negotiations. Neo-liberal approaches are also less effective when there are no level playing fields on which old and new technologies can compete or when existing systems are stabilised by lock-in mechanism. Also the development of radical innovations, which may take decades as this special issue shows, is under-addressed in neo-liberal approaches (which sometimes portray new technologies as 'manna from heaven', i.e. exogenous events). Price signals alone are unlikely to deliver the speed of change required for dealing with climate change (only dramatic price increases may achieve this; but it is politically unfeasible to achieve this through heavy taxation, as the UK fuel protests a few years ago indicate).

(2) *Ecological modernisation* focuses on clean technology. This approach maintains a belief in core modernist principles such as science, technical progress, control, and economic growth. Instead of rejecting modernity, it wants reorientation into more sustainable directions (e.g. Mol 2001). Smart innovations and clean technologies are supposed to create win–win situations: continued economic growth *and* sustainable development. It rejects end-of-pipe solutions which only deal with effects, and shifts attention to sources such as industrial production processes, which need to be redesigned. Examples are process-integrated solutions, reuse and recycling, eco-efficiency, dematerialisation, closing of material loops (as in industrial ecology). Ecological modernisation introduces a welcome 'supply-side' perspective. Many critics, however, question its sufficiency for sustainable development and argue for more radical changes (e.g. Langhelle 2000; York, Van Driel, and Rosa 2003).

(3) '*Deep ecology*' and eco-centrist approaches focus on 'green values' and behavioural change. These approaches argue that environmental problems are fundamentally related to the values of modernity. These values should therefore be rejected and replaced with 'deep green' life styles and localism (Næss 1973; Katz, Rothenberg, and Light 2000). Less radical versions exist, which may propose community-based initiatives where villages or neighbourhoods collectively adopt, maintain and administer 'green' technologies (e.g. biogas plants, solar panels, wind turbines), encouraging each other to change behaviour, roles and responsibilities (see Walker and Devine-Wright 2008, for an overview). While this approach usefully highlights *social* innovations, its radical overtones and (sometimes) technophobia may restrict it to a niche activity.

While these three approaches offer useful elements with regard to problems such as climate change, they are partial and insufficient to bring about the required system changes. They are also fairly abstract, addressing sustainability as a general issue. This special issue elaborates on a fourth position, *socio-technical transitions*, which addresses changes at a more concrete sectoral or systems level (Rochracher 2001; Jacobsson and Bergek 2004; Smith, Stirling, and Berkhout 2005; Geels 2005; Hekkert et al. 2007). This fourth position sees sustainability problems as a formidable societal challenge, whose magnitude is comparable to hygiene and infectious disease problems in nineteenth century 'Western' countries. The latter were addressed through a hygienic transition, which involved technical changes in water supply, sewer systems, housing and food, as well as behavioural changes in washing, bathing, relieving oneself, and cooking, and institutional innovations in local government and service provision (Geels 2006). In analogy, dealing with environmental problems requires shifts to new transport systems, energy systems, food systems, etc. Such transitions not only entail new technologies, but also changes in markets, user practices, policy and cultural discourses, and governing institutions. This fourth position thus aims to overcome the narrow focus of the previous approaches on either markets, technology or behavioural change. Hence, it looks at dynamic interactions and co-evolution between these elements, which as shorthand are labelled 'socio-technical'.

The socio-technical approach focuses on *multiple* actors and social groups, not only firms or consumers/markets. The appropriate analytical level is thus communities or organisational fields, which DiMaggio and Powell (1983, 148) define as:

> those organizations that, in the aggregate, constitute a recognized area of institutional life: key suppliers, resource and product consumers, regulatory agencies, and other organizations that produce similar services or products. The virtue of this unit of analysis is that it directs our attention not simply to competing firms ..., or to networks of organizations that actually interact, ..., but to the totality of relevant actors.

Studied from an organisational field or community perspective (Van de Ven 1993), radical innovation implies not just the emergence of new knowledge and products, but also the creation of a new community, new networks and the rules/institutions that coordinate activities. Several terms have been advanced to emphasise heterogeneity, multi-dimensionality and co-evolution, e.g. 'seamless webs' (Hughes 1986), 'techno-economic networks' (Callon 1991), 'technological systems' (Carlsson and Stankiewicz 1991), 'socio-technical ensembles' (Bijker 1995), and 'techno-institutional complex' (Unruh 2000).

The multi-dimensionality of socio-technical change implies that sustainability transitions are complex processes. This special issue investigates these processes under the heading of 'sustainable innovation journeys'. We use the term 'innovation journey' to capture the open and uncertain nature of radical technological change, which is full of search and exploration processes, twists and turns, etc.[1] The 'journey' aspect also highlights the agency dimension, with actors navigating, negotiating, and struggling their way forward (and sometimes backward as some articles will show). The term also indicates that a transition process can be characterised as a journey of which the final destination is not exactly known. There may be a sense of general direction (towards a sustainable society) but the precise journey changes as the traveller gains more knowledge and experience. We use the term 'sustainable' for two reasons. First, this special issue focuses on the dynamics of *sustainable* technologies, mainly in the energy domain, such as wind, solar and biopower, alternative fuels and low energy housing. Second, innovation journeys need to be *sustained* for long periods of time, often decades rather than years.

With regard to the four policy approaches discussed above, we observe that current sustainability debates are dominated by first and second position. This is unfortunate, in our view, because they are not appropriate for dealing with major environmental problems such as climate change and unlikely to bring about (with sufficient speed) the required structural changes and sustainability transitions. On these topics, innovation studies have important contributions to make to the debate, because of their insights in the dynamics of socio-technical change. These insights also have implications for the kinds of policy instruments that are considered in response to climate change. Economic instruments such as taxes, carbon trading, and R&D subsidies currently dominate. While these instruments *are* important for socio-technical transitions, innovation studies' insights suggest *additional* policies that are related to networks, community building, visions, experiments/learning, etc. Socio-technical approaches refrain from simple policy recipes, because they highlight co-evolution, multi-dimensionality, complexity and multi-actor processes. They argue that constellations of policy instruments should vary, depending on specific challenges, opportunities and problems in sectors, technologies and social networks. While this message may be unpleasant for policy makers who hope for silver bullet solutions, we argue that a deeper understanding of socio-technical dynamics provides policy makers (and other actors) with a more solid base for policy interventions. This special issue aims to contribute to this deeper understanding and to provide insights into effective policy strategies.

Based on these considerations, the papers in this special issue are guided by the following questions:

(1) What are the micro-dynamics of innovation journeys? What are key processes and how should they be conceptualised?
(2) Which policy lessons can be drawn for managing sustainable innovation journeys?

With regard to the first question, we know that multiple disciplinary approaches and conceptual frameworks exist, some of which are used or elaborated in the different papers. Without aspiring to fully answer the first question, Section 2 briefly reflects on the reasons for this variety. That discussion also suggests that each disciplinary approach highlights some key processes and backgrounds others. More sophisticated frameworks therefore tend to arise from *crossovers* between approaches, something that all papers in the special issue practice. Building on this discussion, Section 3 not only introduces the papers, but also provides some tentative ideas on strengths and weaknesses of different frameworks.

The second question is briefly addressed in Section 4. Although we have just emphasised that policies should be tailored to particular contexts, this section articulates some robust policy findings that emerge from the papers in this special issue.

2. Disciplinary approaches and crossovers

Because socio-technical dynamics are multi-dimensional, insights from multiple disciplines are needed to understand the key processes. The following disciplines certainly are relevant in this respect: (1) (industrial and evolutionary) economics and management studies, (2) sociology (especially science and technology studies), (3) political science and (4) cultural studies. These disciplines highlight different analytical elements, ask different questions and open up different 'black boxes', which also means that they may be blind to other aspects.

(1) Economics (especially evolutionary economics and industrial economics) and management studies focus on elements such as: firms, entrepreneurs, competition between old and new technologies, competition between incumbent firms and new entrants, resources (e.g. competencies, money), adaptation and learning through R&D, price/performance developments, industry structures and industrial networks, e.g. user–supplier networks. Broadly speaking then, the basic references are the market and the economy whereas society at large is taken for given. Much attention is given to the *development* of innovation and the formation of new industries. The black box of industrial dynamics is opened up but social embedding is seen as less problematic, or even ignored, and is conceptualised primarily as market adoption (diffusion). What users do with (new) technologies, how they integrate them in user practices, and how they may change their behaviour or preferences, receives less attention.

(2) Sociology (especially science and technology studies) focuses on elements such as: relevant social groups, interpretations and views, motivations, meaning, cognitive frames and (second order) learning, social networks, negotiation and alignment. Broadly speaking, society is the basic reference point, in which the economy is embedded. STS-scholars 'opened up the black box' of technological change by 'following the actors' in their daily practices (e.g. inventors in the laboratory, engineers in design departments). Analysing the co-construction of technology and society, they also look at user contexts and dynamics in application domains (e.g. different user groups, their interpretations, interests, struggles, cultural meanings). But they often pay less attention to *impersonal* mechanisms such as market mechanisms, price/performance competition, resources.
(3) Political science focuses on elements such as: conflicting interests, power struggles, laws, regulations, institutional frameworks, lobbying, advocacy coalitions. Regulations and institutional alignment influence the societal embedding of new technologies while taxes and R&D subsidies influence technological development. Policies are not seen as exogenous aspects that policy makers can influence at will. Instead, political science scholars open up the 'black box of politics', and show the power struggles and conflicts in the shaping of regulatory frameworks. Policy makers are influenced by their constituencies, support in Parliament, re-elections, public opinion, lobby groups, etc.
(4) Cultural studies focuses on elements such as: beliefs, discourses, societal legitimacy, values, social attitudes, social acceptance, cultural enthusiasm. Much attention is given to cultural embedding and social acceptance of new technologies. Culture, however, also influences the supply/development side. For instance, if a new technology has a negative symbolic meaning, this hinders the availability of private and public funding. Culture is not an epiphenomenon, but has real effects. Hence, actors engage in cultural struggles, using metaphors and symbols to influence discourses and shape legitimation processes. Cultural studies, thus, open up the 'black box of culture' underlining that culture is endogenous to sustainable innovation journeys, influencing both the development and societal embedding of new technologies.

The economic and sociological approaches are best represented within innovation studies, as indicated by dedicated professional societies, conferences and professional journals. Evolutionary economics, for instance, has the International Schumpeter Society, the Schumpeter conference, and the *Journal of Evolutionary Economics* as well as the more multidisciplinary *Research Policy*. Likewise, science and technology studies (STS) has the Society for the Social Studies of Science (4S) and The European Association for the Study of Science and Technology (EASST), the 4S and EASST conferences, and journals such *Social Studies of Science* and *Science, Technology and Human Values*. Political science and cultural studies appear to be less institutionalised on these dimensions, at least with regard to the topic of technological change.

The four disciplinary social science approaches each contain a variety of specific theories and perspectives regarding socio-technical change, which conceptualise interactions between relevant analytical elements in different ways. Restricting ourselves to the economic and sociological disciplines, it is not difficult to mention several approaches. Within evolutionary economics some approaches are: self-organisation and complexity theory, path dependence theories, and long wave theories. Some approaches in industrial economics are: industry life cycle approach, the innovation systems tradition, disruptive innovations and technological discontinuities. Within STS some approaches are: actor–network theory, social construction of technology, large technical systems theory.

Table 1. Parent disciplines (+++) and crossovers to other disciplines (+) of the papers in this special issue.

	(Industrial and evolutionary) economics	Sociology (STS)	Political science	Cultural studies
Schot and Geels	+	+++		
Verbong, Geels, and Raven	+	+++		
Bergek, Jacobsson, and Sanden	+++	+	+	
Hillman, Suurs, Hekkert, and Sanden	+++	+	+	
Lovell			+	+++
Agterbosch and Breukers		+	+++	+

While these approaches and theories have recognisable 'parent' disciplines, some of them also make *crossovers* and try to incorporate elements from other disciplines. We want to strengthen a further move to such inter-disciplinary crossovers, because we are more interested in understanding a particular phenomenon (sustainable innovation journeys) than in elaborating disciplinary theories *per se*. Hence, the contributions to this special issue have been selected because they make such crossovers. While the papers in the special issue are 'rooted' in one of the four 'parent' disciplines, they all make crossovers to other disciplines (Table 1). The papers are multi-disciplinary in order to capture the complexity of socio-technical change.

3. Introduction to the papers

The six papers in this special issue focus on different countries (The Netherlands, Sweden, Germany and the United Kingdom) and use different perspectives on innovation journeys. Given the heterogeneity in the disciplinary backgrounds of the contributors, this variety in perspectives is not surprising, but it may be confusing to policy makers and other decision makers seeking ways to influence sustainable innovation journeys. In this section, we will, therefore make a lengthy introduction to the papers, showing how different frameworks conceptualise key processes in innovation journeys.

The first two papers elaborate on and apply the Strategic Niche Management (SNM) approach. The subsequent two papers use the related, yet different approach of Technological Innovation Systems (TIS), with an extension in terms of functional analysis. The final two papers reflect two other approaches: discourse analysis and historical new Institutionalism. These approaches share the view that innovation journeys are complex processes in which several elements and actors become aligned (or not). They also emphasise the ups and downs, and twists and turns of innovation journeys.

SNM initially emphasised the importance of local experimental projects to create 'proto-markets' where users, producers and other stakeholders can interact, learn and exchange experiences. SNM scholars emphasised that new technologies initially need some protection, because they cannot readily compete on mainstream markets (Kemp et al. 1998). Such protection can come from market niches with special selection criteria or from technological niches, such as pilot and demonstration projects. Later literature broadened the scope from single projects to an emerging community (Geels and Raven 2006).

Much SNM-work focuses on an intermediate development phase between R&D and market introduction, where radical innovations need to cross the so-called 'valley of death'. In this relatively early phase, three key processes are distinguished: social network formation, learning processes, and articulation of expectations.[2] Additionally, several sub-processes are identified, especially several dimensions of learning processes. Learning processes are crucial, not just to develop better knowledge and more facts, but also to develop shared meaning. Much attention is, therefore, given to socio-cognitive dynamics, such as conflicting interpretations and a gradual stabilisation of cognitive rules.

The first paper by Schot and Geels outlines the essentials of this influential school of thought. SNM's 'parent discipline' is sociology (STS) but it makes extensive crossovers to evolutionary economics. The paper reviews and discusses the intellectual development of the SNM approach over the past decade, in response to empirical findings and critical comments.

The second paper, by Verbong, Geels and Raven, applies the SNM approach to four renewable energy technologies in the Netherlands: photovoltaics, wind turbines, biomass (digestion, waste incineration, co-firing), and hydrogen (fuel cells). The paper analyses 30-year innovation journeys for these technologies, showing recurrent patterns and policy failures. They conclude that learning processes were too much technology-push, focusing narrowly on R&D, that social networks were closed and supply side oriented and that expectations followed hype-disappointment cycles, partly because policy makers had limited competence to assess promises by product champions.

The third paper, by Bergek, Jacobsson and Sandén, outlines the relatively new extension of technological innovation system (TIS) framework with a 'functions approach'. TIS was initially developed to capture whole transformation processes, including large scale diffusion of new technologies. The initial framework was applied, for instance, to mature technologies and whole sectors, such as factory automation and electronics (Carlsson 1997). Recent contributions in this tradition focus more on early phases in the development of a TIS, distinguishing between key processes at structural and functional levels. Processes at the structural level include entry of organisations, formation of networks and institutional alignment. In this respect TIS builds on industrial economics (and management science) where entry (and exit) patterns have always been central drivers of industrial dynamics, e.g. by influencing division of labour and the formation of legitimacy for new industrial fields.[3]

While its 'parent discipline' is industrial economics (and economics of innovation), the TIS approach attempts to make substantial crossovers to political science and sociology, whereas factors within cultural studies are included in an *ad-hoc* manner only. Crossovers are visible both in attempts to incorporate insights such as political aspects of institutional alignment and in the diverse set of key processes at the functional level, which denotes essential dimensions of what is 'achieved' in the system. Eight dimensions are specified (development of formal knowledge, entrepreneurial experimentation, materialisation, influence on the direction of search, market formation, resource mobilisation, legitimation and development of positive externalities).[4] The paper by Bergek et al. focuses on two of these key processes: 'legitimation' and 'development of positive external economies' (primarily in the Swedish power and transport sectors) and draws implications for both policy makers and managers.

The fourth paper, by Hillman, Suurs, Hekkert and Sandén, applies a variation of the approach 'functions of innovation systems' to the development of alternative fuels in the Netherlands and Sweden. Particular attention is given to the conditions under which cumulative causation takes place. The contrasting outcomes between the Netherlands and Sweden (where technical developments and early market formation have progressed further) add weight to the critique of Dutch policy in Verbong et al.

Both SNM and TIS papers focus broadly on multiple dimensions of innovation journeys. The other two papers (Lovell, Agterbosch and Breukers) focus deeper on specific dimensions.

Lovell's parent discipline is cultural studies, but she makes crossovers to political science. She examines the role of discourse in innovation journeys using the example of low energy housing in the UK (1970–2007). Discourse is shown to unite disparate organisations (which form discourse coalitions), giving structure and direction to innovation journeys. She uses the concept of discursive framing to explain how such coalitions of actors frame sustainable housing innovations as a solution to different policy problems. Such framings are particularly important in the development stage when an innovation needs to gain recognition and visibility among a wider pool of actors. While the paper focuses on one key process (discursive framing), it also highlights associated political and power dimensions.

The empirical analysis shows that an initial framing was made in the context of a social movement of 'radical deep green environmentalism'. A reframing occurred in the 1990s by a low carbon discourse coalition (adopting an ecological modernisation repertoire). This reframing not only accelerated the innovation journey but also reshaped the interpretation of sustainable housing solutions away from the original 'deep ecology' concept, emphasising energy and technology more than waste, water, life styles and other 'softer' issues. The analysis also shows that materialisation through demonstration projects influenced and strengthened the discourse, supporting in a way SNM's emphasis on concrete experimentation and pilot projects.

TIS and SNM can learn from Lovell's in-depth analysis of discourse, which not only influences the framing of problems and opportunities, but also reflects power relations. These insights could provide more depth to the function 'influence the direction of search' in TIS and the key process of 'shaping of expectations' in SNM.

Agterbosch and Breukers' parent discipline is political science but they make crossovers to sociology and cultural studies. They address socio-political processes in innovation journeys, focusing on struggles over interests (e.g. local vs national or global) and cultural meaning (wind turbines as either 'landscape pollution' or symbols of environmental/economic progress). The key process is 'socio-political embedding' which involves the alignment of (formal and informal) institutions and new technology.

The article makes a comparative analysis of the Netherlands and the German state of Nord Rhine Westphalia. The authors apply historical new-institutionalism to assess how the specific institutions in the two 'countries' influenced the socio-political processes of policy generation and implementation at the local level. The analysis emphasises 'institutional alignment', demonstrating in particular how a confluence of policies in energy, spatial planning and environmental domains shaped the implementation process. The paper also highlights the importance of involving locally based investors and grass root movements (as opposed to energy companies) to gain socio-political legitimacy and it demonstrates the inherently political nature of innovation journeys, with actors struggling over interests and cultural meaning. Diffusion and social embedding of new technologies is thus not only a market-driven process, but also entails power struggles and social acceptance, which is influenced by socio-political legitimacy (also the TIS approach distinguishes 'legitimation' as a key process). These insights are also important for the broader political economy of climate change policies, where understanding power relations, and how these may be altered, is critical. Environmental policy making and implementation is not simply managerial. Because power relations are crucial, further analysis of the 'politics of policy' (Jacobsson and Lauber 2006) is called for.

The key processes identified above are internal to an emerging entity (socio-technical ensemble, seamless web, innovation system). The key processes are linked and may strengthen, or weaken, each other. So, positive feedbacks or 'cumulative causation' (Myrdal 1957) are crucial for innovation journeys.

This does not mean that innovation journeys are only driven by endogenous dynamics. External influences from broader contexts are also important. Especially in early phases of development, when innovative novelties are weak and fragile, favourable external contexts (or expectations of it) play important roles. Internal–external interactions are therefore crucial for innovation journeys. More than 50 years ago, Myrdal (1957, 18) formulated the following guiding principle:

> ... the main scientific task is ... to analyse the causal inter-relations within the system itself as it moves under the influence of outside pushes and pulls and the momentum of its own internal processes.

With regard to innovation journeys and transitions, this principle is explicitly elaborated in the multi-level perspective (MLP), in which SNM is contextualised. Distinguishing between regimes and socio-technical landscapes as two kinds of external context, Schot and Geels (2008) explicitly argue that:

> The core notion of the multi-level perspective (MLP) is that transitions come about through interactions between processes at different levels: (a) niche-innovations build up internal momentum, through learning processes, price/performance improvements, and support from powerful groups, (b) changes at the landscape level creates pressure on the regime, (c) destabilisation of the regime creates windows of opportunity for niche-innovations.

In the TIS approach, the distinction between internal and external is, perhaps, less visible, but nonetheless very much there. Indeed, the whole notion of functions was developed to handle an integration of technology-specific and more contextual factors that promote or hinder the development of various functions (Johnson and Jacobsson 2001, 93):

> In the context of an emerging technological system, these factors may be fully technology specific, but may influence several technological systems simultaneously. Hence, they can be derived from a system perspective using different units of analysis: technology, industry, nation.

Hence, driving forces and obstacles to system development are seen as both internal and external (see Section 1 for a discussion of some external obstacles). While TIS and SNM/MLP thus both made a distinction in terms of 'levels', it seems fair to say that the MLP has progressed further in conceptualising interactions between internal and external processes. The energy domain, which is the empirical focus of the special issue, is particularly complex in this respect, because of influences from multiple regimes. For wind power implementation, for instance, Agterbosch and Breukers showed the influence from energy, spatial planning and environmental policy domains. Schot and Geels (2008) therefore propose that multi-niche and multi-regime interactions are important items for future research.

In sum, the papers in this special issue identify a range of key processes. Although the four conceptual frameworks in the papers are multidisciplinary, they still reflect their parent disciplines and therefore highlight specific parts of the 'elephant' whilst playing down or neglecting other parts. To more fully comprehend the dynamics of socio-technical transitions, and support the development of appropriate responses to the climate threat, more crossovers may be needed. While this special issue already demonstrates the fruitfulness of such crossovers, we believe this avenue can be further explored. A clear challenge, from a policy perspective, is to consolidate this richness into a common set of core concepts that can be used to inform policy makers. We hope that future research will continue to synthesise the rich literature on innovation and socio-technical change, and develop multidisciplinary understandings that can inform policy makers, entrepreneurs and other stakeholders who have an interest in sustainable innovation journeys. In the context of climate change policy, there is a particular need to understand better how the process from the initial 'niche' to a large scale transformation can be accelerated. To understand this take-off dynamic, we need to learn more about positive feed-backs between endogenous processes and the influences of external contexts. This is not just a theoretical endeavour, but also a challenge for empirical work and case studies, particularly when regularities, patterns or robust findings can be derived. On that note, the last section of this article describes some robust policy lessons that emerge from the empirical studies in the special issue.

4. Policy lessons

Although the papers differ in their precise conceptual frameworks, we draw seven robust policy lessons from their empirical analyses of sustainable innovation journeys.

First, early market formation is an important process. This is a core idea in both SNM and TIS. Kemp, Schot, and Hoogma (1998) formulate it well:

> Without the presence of a niche, system builders would get nowhere … . Apart from demonstrating the viability of a new technology and providing financial means for further development, *niches help building a constituency behind a new technology*, and set in motion interactive learning processes and institutional adaptation … that are all-important for the wider diffusion and development of the new technology. (Kemp, Schot, and Hoogma 1998, 184, our emphasis)

For radical innovations, new markets thus co-develop with new technologies. This view deviates from the linear model, to which many policy makers still adhere, which sees markets materialise *after* a technology is fully developed. Hillman et al. (2008) show that the Dutch inability to form markets for second generation biofuels hindered the innovation journey, because two technology alliances (the Shell-ECN and the TNO networks) subsequently decided not to build commercial plants. Their contrasting case study of Swedish ethanol and other alternative fuels demonstrates what can happen if early markets *are* formed: investments, materialisation of the technology, and build-up of a constituency further influenced the institutional framework.

Second, consistent and stable policy frameworks are important for innovation journeys, because entrepreneurs need (at least some) stability to make cost/benefit calculations of strategic investments. The papers by Verbong, Geels, and Raven and by Agterbosch and Breukers provide negative cases, where the persistent volatility of Dutch renewable energy policy is a key factor behind failures of many sustainable innovation journeys. A positive example is the relative stability of the German policy regime, centred on the feed-in law and an associated reduction of uncertainties for investors.

Third, social embedding of technology and socio-political legitimacy are important, for both early market formation and further up-scaling. Policy makers often fail to handle this challenge, or only recognise its importance when it is too late (perhaps a consequence of adhering to a linear model). Agterbosch and Breukers, for instance, found such policy failures for wind power (but it may well become a central issue for biofuels as well). Therefore, they conclude that policy makers should address legitimation as a central issue and not as an afterthought to the development and diffusion of new technologies:

> Ex ante evaluation of renewable energy policy with regard to tensions in motives, interests, perceptions and possibilities for different stakeholders at different levels of government is important to foster socio-political embedding. Such an evaluation starts with the acknowledgement of the *political nature of*

> *innovation journeys* and asks for an institutionalisation of participation of relevant stakeholders in policy and planning and decision-making. (our emphasis)

The political and cultural dimensions of social embedding and legitimation imply that technology push policies, with a primary focus on technological learning, have major limitations. Because innovation journeys are co-evolution or co-construction processes, other 'softer' issues need to be handled in parallel with the technological ones. Based on the analysis of four Dutch innovation journeys, Verbong, Geels, and Raven (2008) draw the following critical conclusion:

> All niche-innovation trajectories were characterised by a technology push approach, focussing primarily on R&D. Other dimensions of learning and articulation processes were relatively neglected (e.g. commercial prospects, societal embedding, legal procedures, societal stakeholders). This neglect negatively influenced subsequent implementations. Regarding MCFC, for instance, warnings about the lack of commercial feasibility were consistently ignored, contributing to failure in the first phase. Lack of attention for social acceptance of renewable energies led to major problems with wind turbines and biomass, which both experienced fierce opposition.

Tentative societal experiments with new technologies are one way of testing or articulating social acceptance, especially when many kinds of stakeholders are seriously involved. Lovell (2008) shows that such demonstration or pilot projects may influence broader discourses, which influence legitimacy. Through providing material evidence (of low carbon practices) such projects can provide credence to particular storylines. Lovell therefore emphasised the importance of the first materialisation of a new technology:

> The story lines … have … helped speed up the innovation journey, because their material existence has acted to convince others – government, business and householders – of the economic and technical feasibility of low carbon housing … in other words, the innovative dwellings literally lend substance to the discourse coalition's arguments, and have helped sustainable housing to be viewed as a credible solution to climate change.

Fourth, *multiple* innovation journeys may reinforce each other, especially when they collectively provide legitimacy for certain normative visions (e.g. sustainable development) and build up support for policy changes. Bergek, Jacobsson, and Sandén (2008) therefore argue that 'packs of entrepreneurs' from different innovation journeys may collectively engage in institutional entrepreneurship. The relatedness of such 'packs' may simply consist of supporting the notion of 'renewable energy'. Drawing on case studies of alternative transport fuels in Sweden and renewable power in Germany, they demonstrate that interactions between multiple innovation journeys can generate positive external economies that flow across TIS. Such interactions thus lead to a powerful 'bottom-up' process of institutional alignment and growth. Bergek, Jacobsson, and Sandén, argue that this has implications for management and policy.

> For managers, it becomes essential to align the interests of advocates of several 'competing designs' and, in particular, to work across TIS by forming broader advocacy coalitions. Policy makers need to support such broader coalitions by selecting support schemes that do not pit various emerging TIS against each others in a misguided notion of efficiency. Instead, they should favour regulatory frameworks that stimulate the generation of a diverse set of technological options in parallel.

A fifth issue is the role of incumbents and new entrants in social networks. Large incumbent firms may not be interested in supporting sustainable innovation journeys, because these may threaten their existing products or because they perceive them as uninteresting. This may lead to a lack of investments and difficulties in forming early markets. The technology management literature therefore suggests that new entrants (often small start-up, entrepreneurial firms) are important for developing radical innovations, but we also know that new entrants often fail, because of low legitimacy, lack of political clout, limited resources or insufficient competencies. The incumbent–new entrant issue is one of several policy dilemmas that Schot and Geels (2008) identify. On the one hand, incumbents have many resources and may speed up innovation journeys if they commit themselves. Therefore, policy makers often rely on incumbents to foster new technologies. Agterbosch and Breukers (2008), for instance, show that Dutch policy makers relied on incumbents (e.g. large utilities) to drive wind power until the end of the 1990s. On the other hand, incumbents may have limited motivation or intention to push things forward, as happened in the Dutch wind turbine case. In contrast, the German feed-in law of 1991 explicitly excluded utilities, based on a different economic philosophy that valued new entrants (see Toke and Lauber 2007). Real policy dilemma's have no simple solutions; appropriate answers in one context may not work in another. Policy makers should have sufficient competence to make such assessments, an issue we now turn to.

Sixth, policy makers should have sufficient competence and substantive expertise of the technologies (problems, potential) and the social networks. That way they can better assess the quality of promises from product champions and the motivations of actors (such as the incumbents discussed above). Several papers in the special issue show that this is not always the case. Lovell describes how a low carbon discourse coalition hijacked the innovation journey of low energy housing in the UK, reframing its broad orientation (on waste, water, lifestyles) to a more narrow focus on technology and energy (climate change). The new discourse coalition sidelined the previous network, because it politically and financially more rewarding to frame the innovation journey as 'low carbon'. Policy makers need to be competent enough to not passively go along with such reframing. Bergek, Jacobsson, and Sandén make a similar plea, arguing for a competence base that permits decision makers to critically assess attempts to shape expectations and normative legitimacy, in particular by incumbents with larger resources and greater access to media. Verbong, Geels, and Raven also identify limited competence and poor policy learning as a key factor behind Dutch failures in sustainable innovation journeys. They identify hype-cycles in Dutch policy, caused by policy makers uncritically accepting promises of product champions, followed by disappointments when setbacks occur.

Seventh, because sustainable innovation journeys are multifaceted, policies need not be limited to R&D subsidies and changing relative prices (as economists and model building engineers often think). Innovation journeys are shaped by not only economic, but also social, political and cultural factors. Policy makers should acknowledge this multifaceted nature of innovation journeys, and use a variety of policy instruments. From the relative success of Swedish alternative fuels, Hillman et al. draw the lesson that a broad policy orientation is needed, which addresses multiple functions and, thereby contributing to build a complete TIS:

> ... policy measures are most effective when designed to target all the TIS functions. Also entrepreneurs are most successful in bringing about change when actions are directed at facilitating multiple system functions; this is actually what makes *cumulative causation possible* .... The possible advantages of a broader policy orientation is shown by the variety of actions taken by the national Swedish government. Besides guidance, Swedish policy targeted knowledge formation, but also ... early entrepreneurial experiments and the (niche) market activities around 1G biofuels .... (Our emphasis)

The challenge is, thus, to avoid the temptations of the 'linear model' and adopt a broader conceptualisation that addresses the multi-dimensionality of technological niches, technological innovations systems or other terms that analysts use to highlight the intrinsic interrelatedness of socio-technical change (e.g. seamless webs, socio-technical ensembles). We hope this special issue has contributed to developing this broader conceptualisation.

Acknowledgements

The papers in this special issue were presented and discussed at a workshop in Utrecht, in October 2006. We are very grateful not only to the participants, but also to the sponsors, VINNOVA and KSI (Knowledge network on System Innovation), for making this workshop a success.

Notes

1. The term 'innovation journey' has been defined by van de Ven et al. (1999, 6–7) as 'new *ideas* that are developed and implemented to achieve desired *outcomes* by *people* who engage in *transactions* (relationships) with others in changing institutional and organizational *contexts*' (italics in original).Where van de Ven et al. (1999) focused on innovation journeys within organisations, we will focus on sustainable innovation journeys at the level of societies, sectors and nations.
2. Because of its focus on relatively early phases, SNM is less explicit about further diffusion. More attention could be given to the subsequent 'take-off' phase. As the stock of a particular technology increases 100 or even 500 times, we may need to know more about the mobilisation of resources (money, competencies). This might require more understanding of the capital market and adjustments in the educational system to 'produce' the required skills. Also the entry of new organisations seems an important process for take-off, because of its influence on resource mobilisation, division of labour (specialised suppliers), scale economies, etc. In our view, SNM might benefit from insights in industrial economics, where a rich literature has addressed these kinds of issues (e.g. Smith, Marx, Marshall, Young, Rosenberg).
3. Because of different disciplinary backgrounds, 'functions in innovation systems' has a more economics/entrepreneurial flavour and SNM a more socio-cognitive one.
4. Hillman et al. (2008) uses seven functions, where some differ from those used by Bergek, Jacobsson, and Sandén (2008).

References

Agterbosch, S., and Breukers, S. 2008. Socio-political embedding of onshore wind power in the Netherlands and North Rhine-Westphalia. *Technology Analysis & Strategic Management* 20, no. 5: 633–48.

Bergek, A., S. Jacobsson, and B.A. Sandén. 2008. 'Legitimation' and 'development of external economies': two key processes in the formation phase of technological innovation systems. *Technology Analysis & Strategic Management* 20, no. 5: 633–48.

Bijker, W.E. 1995. *Of bicycles, bakelites and bulbs: towards a theory of sociotechnical change*. Cambridge, MA: The MIT Press.

Callon, M. 1991. Techno-economic networks and irreversibility. In *A sociology of monsters: essays on power, technology and domination*, ed. J. Law, 132–61. London: Routledge.

Carlsson, B., and R. Stankiewicz. 1991. On the nature, function and composition of technological systems. *Journal of Evolutionary Economics* 2, no 1: 93–118.

Carlsson, B., and S. Jacobsson. 1997. Diversity creation and technological systems: a technology policy perspective. In *Systems of innovation: technologies, institutions and organizations*, ed. C. Edquist. London: Pinter.

Carlsson, B., ed. 1997. *Technological systems and industrial dynamics*. Boston, MA: Kluwer Academic.

DiMaggio, P.J., and W.W. Powell. 1983. The iron cage revisited: institutional isomorphism and collective rationality in organizational fields. *American Sociological Review* 48: 147–60.

Freeman, C., and C. Perez. 1988. Structural crisis of adjustment, business cycles and investment behaviour. In *Technical change and economic theory*, eds. G. Dosi, C. Freeman, R. Nelson, G. Silverberg and L. Soete, 38–66. London: Pinter.

Geels, F.W. 2002. Technological transitions as evolutionary reconfiguration processes: a multi-level perspective and a case-study. *Research Policy* 31, no. 8–9: 1257–74.

———. 2005. *Technological transitions and system innovations: a co-evolutionary and socio-technical analysis*. Cheltenham: Edward Elgar.

———. 2006. The hygienic transition from cesspools to sewer systems (1840–1930): the dynamics of regime transformation. *Research Policy* 35, no. 7: 1069–82.

Geels, F.W., and R.P.J.M. Raven. 2006. Non-linearity and expectations in niche-development trajectories: up and downs in Dutch biogas development (1973–2003). *Technology Analysis & Strategic Management* 18, no. 3–4: 375–92.

Hekkert, M.P., R.A.A. Suurs, S.O. Negro, S. Kuhlmann, and R.E.H.M. Smits. 2007. Functions of innovation systems: a new approach for analysing technological change. *Technological Forecasting and Social Change* 74, no. 4: 413–32.

Hillman, K.M., R.A.A. Suurs, M.P. Hekkert, and B.A. Sandén. 2008. Cumulative causation in biofuels development: a critical comparison of the Netherlands and Sweden. *Technology Analysis & Strategic Management* 20, no. 5: 593–612.

Hughes, T.P. 1986. The seamless web: technology, science, etcetera, etcetera. *Social Studies of Science* 16, no. 2: 281–92.

Jacobsson, S., and V. Lauber. 2006. The politics and policy of energy system transformation – explaining the German diffusion of renewable energy technology. *Energy Policy* 34, no. 3: 256–76.

Jacobsson, S., and A. Bergek. 2004. Transforming the energy sector: the evolution of technological systems in renewable energy. *Industrial and Corporate Change* 13, no. 5: 815–49.

Johnson, A., and S. Jacobsson. 2001. Inducement and blocking mechanisms in the development of a new industry: the case of renewable energy technology in Sweden. In *Technology and the market, demand, users and innovation*, eds. R. Coombs, K. Green, A. Richards and W. Walsh. Cheltenham: Edward Elgar.

Katz, E., D. Rothenberg, and A. Light, eds. 2000. *Beneath the surface: critical essays in the philosophy of deep ecology*. Cambridge, MA: MIT Press.

Kemp, R., J. Schot, and R. Hoogma. 1998. Regime shifts to sustainability through processes of niche formation. The approach of strategic niche management. *Technology Analysis and Strategic Management* 10, no. 2: 175–95.

Langhelle, O. 2000. Why ecological modernization and sustainable development should not be conflated. *Journal of Environmental Policy & Planning* 2: 303–22.

Lovell, H. 2008. Discourse and innovation journeys: the case of low energy housing in the UK. *Technology Analysis & Strategic Management* 20, no. 5: 613–32.

Mol, A.P.J. 2001. *Globalization and environmental reform: the ecological modernization of the global economy*. Cambridge, MA: MIT Press.

Myrdal, G. 1957. *Economic theory and underdeveloped regions*. London: Ducksworth.

Næss, A. 1973. The shallow and the deep, long-range ecology movement. *Inquiry* 16: 95–100.

Rohracher, H. 2001. Managing the technological transition to sustainable construction of buildings: a socio-technical perspective. *Technology Analysis & Strategic Management* 13, no. 1: 137–50.

Schot, J., and Geels, F.W. 2008. Strategic niche management and sustainable innovation journeys: theory, findings, research agenda, and policy. *Technology Analysis & Strategic Management* 20, no. 5: 537–54.

Schumpeter, J.A. 1934. *The theory of economic development: an inquiry into profits, capital, credit, interest, and the business cycle*. Cambridge, MA: Harvard University Press.

Smith, A., A. Stirling, and F. Berkhout. 2005. The governance of sustainable socio-technical transitions. *Research Policy* 34: 1491–510.

Stern, N. 2006. *Stern review on the economics of climate change*. London: H.M. Treasury, United Kingdom.

Toke, D., and V. Lauber. 2007. Anglo-Saxon and German approaches to neoliberalism and environmental policy: the case of financing renewable energy. *Geoforum* 38: 677–87.

Unruh, G.C. 2000. Understanding carbon lock-in. *Energy Policy* 28: 817–30.

Van de Ven, A.H. 1993. A community perspective on the emergence of innovations. *Journal of Engineering and Technology Management* 10: 23–51.

Van de Ven, A.H., D.E. Polley, G. Garud, and S. Venkataraman. 1999. *The innovation journey*. Oxford: Oxford University Press.

Verbong, G., F.W. Geels, and R. Raven. 2008. Multi-niche analysis of dynamics and policies in Dutch renewable energy innovation journeys (1970–2006): hype-cycles, closed networks and technology-focused learning. *Technology Analysis & Strategic Management* 20, no. 5: 555–73.

Walker, W. 2000. Entrapment in large technology systems: institutional commitments and power relations. *Research Policy* 29, no. 7–8: 833–46.

Walker, G.P., and P. Devine-Wright. 2008. Community renewable energy: what should it mean? *Energy Policy* 36: 497–500.

York, R., H. Van Driel, and E.A. Rosa. 2003. Key challenges to ecological modernization theory. *Organization & Environment* 16, no. 3: 273–88.

Frank W. Geels
Department of Technology Management
Eindhoven University
The Netherlands

Marko P. Hekkert
Copernicus Institute for Sustainable Development and Innovation
Universiteit Utrecht
The Netherlands

Staffan Jacobsson
Environmental Systems Analysis, Department of Energy and Environment
Chalmers University of Technology
Gothenburg, Sweden

Strategic niche management and sustainable innovation journeys: theory, findings, research agenda, and policy

Johan Schot and Frank W. Geels

Eindhoven University of Technology, The Netherlands

This article discusses empirical findings and conceptual elaborations of the last 10 years in strategic niche management research (SNM). The SNM approach suggests that sustainable innovation journeys can be facilitated by creating technological niches, i.e. protected spaces that allow the experimentation with the co-evolution of technology, user practices, and regulatory structures. The assumption was that if such niches were constructed appropriately, they would act as building blocks for broader societal changes towards sustainable development. The article shows how concepts and ideas have evolved over time and new complexities were introduced. Research focused on the role of various niche-internal processes such as learning, networking, visioning and the relationship between local projects and global rule sets that guide actor behaviour. The empirical findings showed that the analysis of these niche-internal dimensions needed to be complemented with attention to niche external processes. In this respect, the multi-level perspective proved useful for contextualising SNM. This contextualisation led to modifications in claims about the dynamics of sustainable innovation journeys. Niches are to be perceived as crucial for bringing about regime shifts, but they cannot do this on their own. Linkages with ongoing external processes are also important. Although substantial insights have been gained, the SNM approach is still an unfinished research programme. We identify various promising research directions, as well as policy implications.

1. Introduction

Following Mokyr, we can describe new technologies as 'hopeful monstrosities' (Mokyr 1990, 291). They are 'hopeful', because product champions believe in a promising future, but 'monstrous' because they perform crudely. As Rosenberg (1976, 195) argues: 'most inventions are relatively crude and inefficient at the date when they are first recognised as constituting a new invention. They are, of necessity, badly adapted to many of the ultimate uses to which they will eventually be put.' This means that new technologies cannot immediately compete on the market against established technologies. This problem is pivotal for many new technologies with sustainability promise for energy, transportation, agriculture, etc. There is no lack of such new technologies, which are developed in R&D laboratories and put to use in demonstration projects. They have a hard time, however, bridging the 'valley of death' between R&D and market introduction. Building on Van de Belt and Rip (1984), Schot, Hoogma, and Elzen (1994) and Kemp, Schot, and Hoogma (1998) introduced the strategic niche management perspective (SNM) to address this problem.[1] Hence, much of the SNM literature focuses on understanding the early adoption of new technologies with high potential to contribute to sustainable development.

A core assumption of the SNM approach is that sustainable innovation journeys can be facilitated by modulating of technological niches, i.e. protected spaces that allow nurturing and experimentation with the co-evolution of technology, user practices, and regulatory structures. Radical innovations therefore never serve as a simple technological fix in an SNM approach. It is not a technology push approach. Instead, SNM scholars argue that sustainable development requires interrelated social and technical change. In that respect, they build on the work of sociologists of technology who argue that technological and social change are interrelated.[2] They also recognise that the rise of modernity created conditions in which technology actors usually focus on developing, testing and optimising technology, but neglect the embedding in broader societal goals, or leave it to a later stage. Confronted with relatively fixed technical designs, governments often take on the role of mitigating and/or compensating for the impacts. This historical diagnosis has been the background for the development of new thinking about technology policy and Constructive Technology Assessment of which SNM is an offspring.[3]

While early SNM scholars had normative concerns, subsequent research did not focus primarily on the development of SNM as a policy tool, although some work has been done in that direction. Instead, SNM scholars began investigating which processes determine successful niche development. This research direction was based on the idea that strategies such as SNM can only work well when they modulate on-going dynamics. Consequently, SNM as a policy tool does not suggest that governments create niches in a top–down fashion, as is sometimes assumed by commentators, but focuses instead on endogenous steering, or steering from within.[4] Such steering can be enacted by a range of actors, including users and societal groups. Steering can address many parts of the process, by adding a new actor, a specific learning process or a set of demonstration projects which may redirect evolving dynamics toward a desired path. Niches are not inserted by governments, but are assumed to emerge through collective enactment. Nevertheless, their (future) course can be modulated into more sustainable directions. Because of these characteristics, we would like to define SNM as a form of reflexive governance (see Voss, Bauknecht, and Kemp 2006; and in particular, see Grin 2006).

During the last 10 years, many new studies and articles have appeared, which have reported empirical findings and suggested further elaboration. In this article, we review and discuss the results of this research and show how the research agenda has evolved. The article is structured as follows. In Section 2 we provide the theoretical background of SNM and introduce some basic concepts. In particular we discuss a specific conceptualisation of the role of niches in technical change. In Section 3 we discuss the main body of SNM research that focuses on niche-internal processes. In Section 4 we broaden our scope and look at conceptualisations that connect niche-internal to external processes. In Section 5 we address policy implications of SNM research. Finally, the concluding section brings together the results of this discussion and articulates a future research agenda. Throughout the article we distinguish between 'early' SNM research which mainly focuses on niche-internal processes and 'later' SNM research which provides a more elaborate analysis of the interaction between niches and their broader environments.

2. Theoretical background of SNM research

The theoretical background of SNM consists of an attempt to import insights from constructivist science and technology studies into evolutionary economics as developed by Nelson and Winter (1982) and Dosi (1982). This effort led to the formulation of a quasi-evolutionary perspective on technical change by Rip (1992, 1995) and Schot (1992, 1998). They argue that variation is not blind, as is assumed by many evolutionary economists, but directed to some extent. Technology actors not only anticipate future selection, but also try to shape the selection process itself by setting up special programmes in R&D settings or demonstration projects. These are spaces in which radical novelties are tried out and developed further, while they are sheltered from mainstream competition. This way, firms and governments develop variations, which do not (yet) fit in the existing selection environment. Rip and Schot referred to such spaces as (technological) niches.[5] In their view, these niches could be used not only for testing the design of the technology, but also for mutual articulation (and alignment) of technology, demand and broader societal issues, including sustainable development. The niche provides a protected space that nurtures a specific set of interactions between issues, but also between actors representing these issues. The additional hypothesis was that such niches function as 'proto-markets' which may jumpstart the development of market niches (i.e. niches in which technology design and user demands have become stabilised). Technological niches often involve policy makers, but they do not do so necessarily. Users and societal groups can also invest their time and resources to construct niches.[6] One important reason for governments to subsidise and nurture not yet profitable innovations is the expectation that they will become important for realising particular societal and collective goals in the future.[7] Because of these expectations, governments and other actors operating in the niche may accept disadvantages in the present and invest resources in upgrading and developing a 'hopeful monstrosity'.

The notion of 'niche' was already present in the innovation literature but was not focused on inducing sustainable development. Evolutionary economists (Saviotti 1996; Windrum and Birchenhall 1998; Frenken, Saviotti, and Trommetter 1999) and management scholars have emphasised the importance of market niches for radical innovations (Astley 1985; Lynn, Morone, and Paulson 1996). In particular, Levinthal's (1998) path-breaking article argues that radical change may occur as a result of distinct selection criteria operating in a niche. In his conceptualisation the initial speciation event is minor in the sense that the technology does not differ substantially from its predecessors, but different selection pressures in the niche subsequently trigger a divergent evolutionary path. His idea of a market niche thus rests on the assumption that selection environments are not homogenous and consist of several different niches. While Levinthal and others usually take the existence of market niches for granted, and assume that minor variations of prevailing technologies are tried out in such niches, SNM scholars argue that for many innovations, especially with sustainability promise, market niches and user demand are not readily available because the innovations are not minor variations from the prevailing set of technologies, but differ radically from them.[8] SNM was thus developed to serve the management of a particular type of innovations: (1) socially desirable innovations serving long-term goals such as sustainability, (2) radical novelties that face a mismatch with regard to existing infrastructure, user practices, regulations, etc. It is precisely for this reason that SNM scholars see real-world experimental projects as important devices that precede market niche development. Such proto-markets can be exploited to explore possible alignments of technology, user demands and sustainability issues.

3. Niche internal processes

In early SNM work the idea is that the selective exposure of new (sustainable) technologies to the market through a process of niche development can eventually lead to the replacement of the dominant (polluting) technologies. This replacement would take the form of the development of a new socio-technical regime that carries and stores the rules (partly embodied in standards, skills, designs, and government regulations) for how to produce, use and regulate the new technology. Early SNM work conceptualised the process as a bottom–up process, in which novelties emerge in technological niches, then conquer market niches, and eventually replace and transform the regime (Figure 1).

The main research question was: how and under what circumstances is the successful emergence of a technological niche possible? Success was defined in terms of transformation of a technological niche into a market niche and eventually a regime shift. Based on a range of insights from innovation studies, including STS, evolutionary economics and history of technology, three (internal) processes were distinguished for successful development of a technological niche (Elzen, Hoogma, and Schot 1996; see also Kemp, Schot, and Hoogma 1998):[9]

(1) The articulation of expectations and visions. Expectations are considered crucial for niche development because they provide direction to learning processes, attract attention, and legitimate (continuing) protection and nurturing.
(2) The building of social networks. This process is important to create a constituency behind the new technology, facilitate interactions between relevant stakeholders, and provide the necessary resources (money, people, expertise).
(3) Learning processes at multiple dimensions:

   (a) technical aspects and design specifications
   (b) market and user preferences
   (c) cultural and symbolic meaning
   (d) infrastructure and maintenance networks
   (e) industry and production networks
   (f) regulations and government policy
   (g) societal and environmental effects

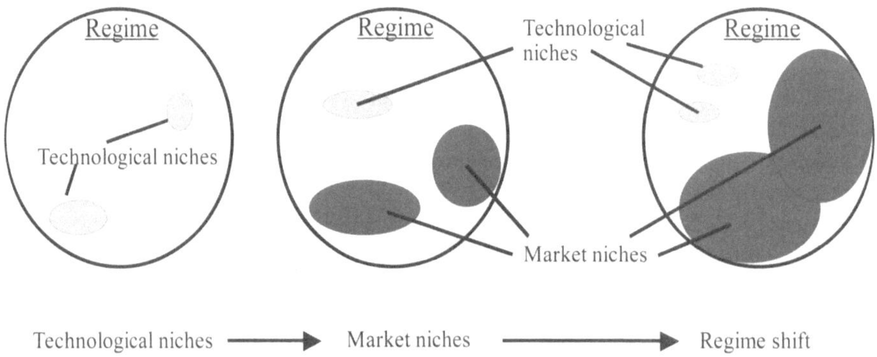

Figure 1. From niche dynamics to regime shift (adapted from Weber et al. 1999, 22).

Subsequently, more specific hypotheses were formulated for each process (Elzen, Hoogma, and Schot 1996, 76–78; and Hoogma et al. 2002, 28–29):

(1) Expectations would contribute to successful niche building if expectations were made: (a) more robust (shared by more actors), (b) more specific (if expectations are too general they do not give guidance), and (c) have higher quality (the content of expectations is substantiated by ongoing projects);
(2) Social networks are likely to contribute more to niche development if: (a) the networks are broad, i.e. multiple kinds of stakeholders are included to facilitate the articulation of multiple views and voices; the involvement of relative outsiders may be particularly important to broaden cognitive frames and facilitate second-order learning; (b) the networks are deep, i.e. people who represent organisations, should be able to mobilise commitment and resources within their own organisations and networks;
(3) Learning processes would contribute more to niche development if they are not only directed at the accumulation of facts and data, i.e. first-order learning, but also enable changes in cognitive frames and assumptions, i.e. second-order learning (derived from Grin and Van de Graaf 1996).

These hypotheses were tested in a European Union project,[10] three PhD theses (Hoogma 2002; Van Mierlo 2002; Raven 2005) and discussed, criticised or amended in several other studies.[11] These studies contained empirical (case) studies of finished and/or ongoing experiments in a range of fields, from transport to energy to agriculture and sanitation, mainly in (Western) European contexts, but also in Tanzania and South Africa. They investigated if the identified success conditions explained outcomes. The case selection included some examples of market niche development, but many cases featured a limited outcome in terms of inducing further niche development into a sustainable direction.

The results showed that many demonstration projects were organised in an overly contained way. Networks tended to be narrow and projects tended to focus on first-order leaning. Consequently, many demonstration projects followed too much of a technology push approach. The narrow focus came through in the way users were included in the demonstration projects that were studied. They were mainly perceived as consumers with given needs and preferences. Hence, the aim of many demonstration projects was to discover (mis)matches between technology features and these (assumed) needs. Standardised surveys and usability trials and panels were used to investigate these (mis)matches. Failed niche developments could often be related to either minimal involvement of outsiders in the experiments and a lack of second order learning, or to minimal involvement of regime actors which resulted in lack of resources and institutional embedding. Another recurring finding is that the nature of social networks determined the depth and breadth of learning processes. Networks that were broad and contained outsiders provoked more second-order learning. These studies show that SNM is a useful *ex-post* analytical framework. The demonstration projects in these studies did not use SNM prescriptively as a management tool. So the real-life problems in these projects are not indicative for theoretical failures in SNM, as some commentators have suggested.

Some of these studies highlight shortcomings of the SNM approach as defined in Kemp, Schot, and Hoogma (1998) and Hoogma et al. (2002). For instance, Brown et al. (2004) and Harborne et al. (2007) stress that involvement of outside actors and second-order learning do not happen easily and by themselves. It requires the presence of particular drivers and contexts. They point to the importance of a sense of urgency and the role that a process of structured repeated visioning could play. In similar vein, Hegger, van Vliet, and B.J.M. van Vliet (2007) argue that the strong focus on experiments with technology in many demonstration projects is not conducive to broad learning and outsider involvement. It might reinforce the technology push character of actual experiments. They propose to redirect the focus of niche experiments towards concepts, visions and guiding principles rather than technologies, and toward experimenting with social aspects first, albeit without neglecting the socio-technical character of the change process. Finally, the transition management (TM) approach advocated by Rotmans, Loorbach and others, which like SNM highlights the importance of experiments, emphasises the importance of creating visions before starting experiments (Rotmans, Kemp, and van Asselt 2001; Loorbach 2007; Kemp, Rotmans, and Loorbach 2007).[12] TM promotes envisioning practices in so-called 'transition arenas', which consist of regime actors, niche actors and outsiders. TM thus actively aims to influence the regime, using niche experiences and alternative visions to influence the cognitive frames of regime actors. Grin (2006) makes similar points about the transformative power of influencing cognitive frames, which he sees as an important aspect of reflexive governance (see also Bos and Grin 2008, forthcoming). Grin argues that biases and limitations in existing institutions can be overcome by providing actors with a meta-vision that helps them deal with the challenge of creating fundamental change.

On the one hand, we acknowledge that TM addresses some factors that SNM underplays. While SNM develops an evolutionary approach that builds on and leverages the dynamic forces of market competition, aimed at overcoming lock-in and promoting socio-technical diversity, TM suggests a more ambitious approach of goal-oriented modulation that places more emphasis on the role of strategic envisioning. In that respect, TM introduces the notion of 'transition experiment' which is supposed to be different from regular innovation experiments (Van den Bosch and Taanman 1996).

On the other hand, in practice there are too many fruitless scenarios and visioning exercises, with few substantive follow-up activities. In a critical interpretation, one might say that many of these exercises have become rituals, where actors express good intentions as a form of public 'impression management'. While we recognise that reproductions of rituals may sometimes provide conditions for change, there are many instances where they have little real influence. Hence, SNM scholars have stressed the importance of 'hands-on', real-life experiences in demonstration projects; SNM assumes that actual implementation and specification of visions in experimental settings is most conducive for niche development. We are not yet convinced that there is much to gain from visioning beforehand in transition arenas. Still, we also recognise that SNM's assumptions may need to be reconsidered: empirical research of sequences of experimental projects indicates that visions and expectations do not evolve as much as we expected in response to learning processes in the projects. Several critical sympathisers (e.g. Hegger, Harborne, and Brown 2007) have argued that visioning prior to experimentation does help to broaden networks and learning processes.

In our opinion, this controversy touches upon a central problem of technology development in modern societies. Technology actors such as firms and governments introducing new technologies tend to exclude certain actors and focus on optimising the technological side first while neglecting other social aspects. It remains to be seen whether introducing some structured process of visioning (as in transition management) in arenas arranged by policy actors, or forms of conceptual niche management as proposed by Hegger, Harborne, and Brown (2007), could indeed help to overcome this modernist bias. Early SNM put its cards on influencing the actual design and implementation of a range of new varieties. Below, we show that later SNM suggests that such a strategy needs to complemented with other measures which modulate emerging windows of opportunity external to the niche.

A lot of the cited research focuses on explaining the limited success of the experiments studied. Conclusions point to the conditions that would better encourage particular types of learning, networking and visioning. Hommels, Peters, and Bijker (2007) have argued that part of the problem might be that SNM focuses too much on providing protection. In their view innovations have a better chance of success if made vulnerable by subjecting them to risks and oppositions from the outset. They developed a management tool (PROTEE) to arrange for learning process about the context in which the innovation will be embedded. This tool might indeed be useful for an SNM approach. Contrary to what Hommels, Peters, and Bijker (2007) argue, controlled exposure to selection pressures has been central to SNM research from the start. Yet we agree that more attention should be devoted to ways in which protection is provided and can be lifted in a phased way. The managing of selective pressures is not only an issue of specific measures, such as subsidies, but also one of niche expansion and the emergence of a new set of stable rules and routines.[13] Yet, innovations in SNM are of a particular nature (see above) that requires some form of protection. Otherwise the journey would not begin at all since market demand does not pull and firms and other technology actors are not pushing for market introduction, as Harborne, Hendry, and Brown (2007, 184) formulated the problem.

Many of the studies discussed can be seen as inquiries into understanding the failure and successes of the journey from technological niche to market niche, and eventually to a regime shift. On this point, Raven (2005) and Van Mierlo (2002) made a crucial contribution by signalling the need for distinguishing between local socio-technical projects and the niche level which consists of an emerging community that shares cognitive, formal and normative rules (Figure 2).

Niche development can then be conceptualised as progressing at two levels simultaneously: the level of projects in local practices and the global niche level. Sequences of local projects may gradually add up to an emerging field (niche) at the global level (Figure 3). Using the work of Deuten (2003), Geels and Raven (2006) conceptualised this aggregation process as follows: developments may start with one or a few projects, carried by local networks of actors, who are interested in innovations for idiosyncratic or local reasons. The cognitive rules (such as expectations) that guide these projects are initially diffuse, broad and unstable. Local projects form test beds for these diffuse ideas and spaces for the elaboration of new ideas. If learning processes in local projects are compared and aggregated, the cognitive rules at the more global niche level may gradually become more articulated, specific and stable. In this conceptualisation, a technological niche is not only characterised by protection (which tends to be phased out slowly), but also by the locality and instability of rules and networks. The movement to a market niche does not only entail a movement to more exposure to selection pressures, but also to more stable shared rules (e.g. dominant designs).

This conceptualisation shifts the attention from single projects and their success or failure to sequences of projects, which can accumulate into learning trajectories, while also the notion

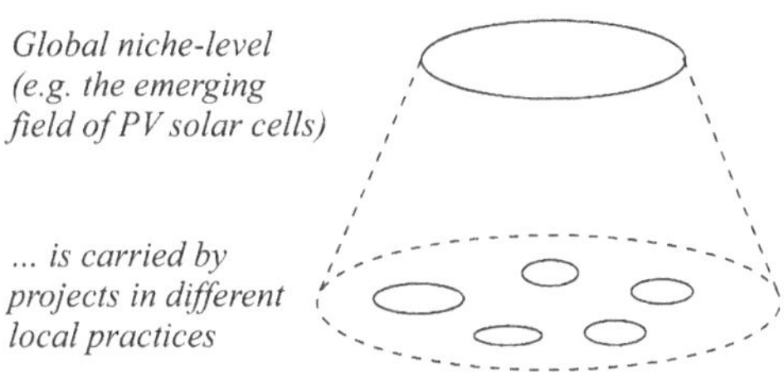

Figure 2. Local projects and global niche-level (Geels and Raven 2006, 378).

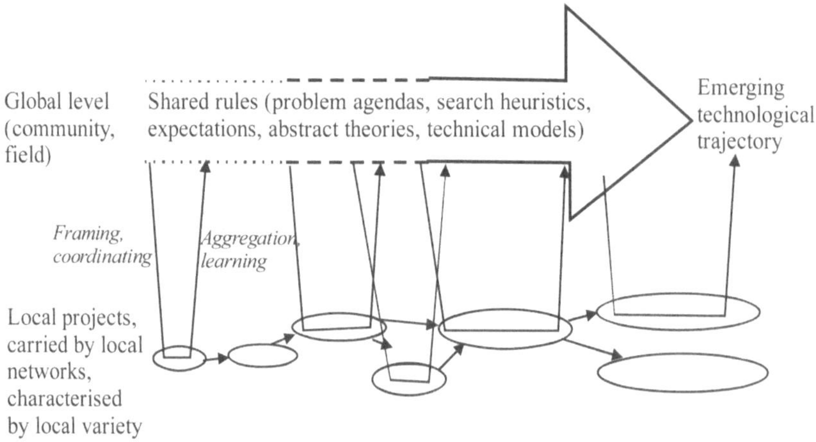

Figure 3. Emerging technical trajectory carried by local projects (Geels and Raven 2006, 379).

of failure itself becomes more layered since failed projects can contribute to the success of the overall sequence. This point is reinforced by Van den Bosch and Taanman (1996) who discuss the importance of a cyclical pattern of learning and networking that would help to create a set of more global rules, and by Van Eijck and Romijn (2008) who stress the importance of organising sequences which take into account changes needed in the entire production chain. This line of research opens up a range of new topics and questions that require further investigation, in particular in two areas: (1) mechanisms and factors that make sequences of projects gel into niche development; (2) contributions of interactions among between multiple regimes and niches towards regime shifts.

These areas have been explored in some recent papers. Geels and Raven (2006) argue that sequences of projects are guided by cognitive rules and expectations, thus restating the importance of visions, albeit not for developments within projects but between projects. They also point to changes in external circumstances such as oil prices and the liberalisation of the electricity sector that influence the adoption and direction of developments. Geels and Deuten (2006) emphasise the role of intermediary actors at the community level (e.g. branch organisations, professional societies), who monitor multiple local projects, aggregate generic lessons, and circulate knowledge through journals or dedicated workshops and conferences. Earlier, Van Mierlo (2002) found that Shell was involved in most projects with solar photovoltaics in the Netherlands in the late 1990s. This professional actor brought lessons from one project along to the next project. Different projects, however, also compete with other, so actors may not be willing to share learning experiences. Secrecy may hamper circulation of lessons and experiences. This issue of competition leads to a bigger issue. SNM assumes that diversity is productive for niche development, because it enhances learning and network development, but too much diversity may hamper developments, because it creates uncertainty (which prevents full commitments), fragments resources and hampers the emergence of a stable set of rules. This dilemma needs more attention in future research.

These findings and discussions suggest that the journey from experiments to regime shift is more complicated than was previously assumed. In 2002, Hoogma et al. acknowledged this:

> For one thing, we were certainly over-optimistic about the potential of SNM as a tool for transition . . . . The positive circles of feedback by which a technology comes into its own and escapes a technological niche, are far weaker than expected and appear to take longer than expected (5 years or more) . . . . The

> experiments did not make actors change their strategies and invest in the further major development of a technology . . . . The experiments were relatively isolated events. It seems difficult for the actors to build bridges. Although more could perhaps have been done and achieved, there are limits to the power of experiments. Only occasionally will an experiment be such a big success that it will influence strategic decisions. Experiments may tip the balance of decision-making, but they will not change the world in a direct, visible way . . . . Experiments influence the world but do not bring particular futures about. Their influence is more indirect. (Hoogma et al. 2002, 195–196)

In sum, SNM has identified and empirically investigated important niche-internal mechanisms in sustainable innovation journeys. While SNM research provides evidence that there is a correlation between the design of experiments and outcomes in terms of technological and market niche development, it is also clear that internal niche developments are not the only important factor. External factors also play a crucial role. Niche innovations are rarely able to bring about regime transformation without the help of broader forces and processes. This conclusion led to a search for conceptualisations that linked niche internal and external processes. This search was done under the heading of the multi-level perspective, and developed in parallel with much of the SNM work discussed above.

4. Niches and multi-level analysis

A broader, contextualised view of the role of niches in technical change was developed in the context of a number of research projects, in particular: (a) a literature review and synthesis about technological change, performed for the Batelle Memorial Institute funded by the US Department of Energy (Rip and Kemp 1998); (b) a major research programme about the history of technology in the Netherlands in the nineteenth and twentieth century,[14] and (c) PhD research by Geels (2002, 2005) into technological transitions and regime shifts, based on historical case studies. This work culminated in a multi-level perspective (first formulated by Rip and Kemp 1998) that distinguishes three analytical levels. Niches form the micro-level where radical novelties emerge. The socio-technical regime forms the meso-level, which accounts for the stability of existing large-scale systems (in transport, energy etc). The socio-technical regime is an extended version of Nelson and Winter's (1982) technological regime, which referred to shared cognitive routines in an engineering community and explained patterned development along 'technological trajectories'. Sociologists of technology broadened this explanation, arguing that scientists, policy makers, users, and special-interest groups also contribute to patterning of technological development (e.g. Bijker 1995). The socio-technical regime concept accommodates this broader community of social groups and the alignment of activities. Regimes not only refer to cognitive routines and belief systems, but also to regulative rules and normative roles. The macro-level is formed by the socio-technical landscape, an exogenous environment beyond the direct influence of niche and regime actors (e.g. macro-economics, deep cultural patterns, macro-political developments). Changes at the landscape level usually take place slowly, in the order of decades.[15] Figure 4 indicates that the three levels form a nested hierarchy with regard to local practices. Niche actors hope that novelties will eventually be used in the regime or even replace it. This is not easy, because the existing regime is stabilised and entrenched in many ways (lock-in).

The core notion of the multi-level perspective (MLP) is that transitions come about through interactions between processes at different levels: (a) niche innovations build up internal momentum, (b) changes at the landscape level create pressure on the regime, (c) destabilisation of the regime creates windows of opportunity for niche innovations (Figure 5).

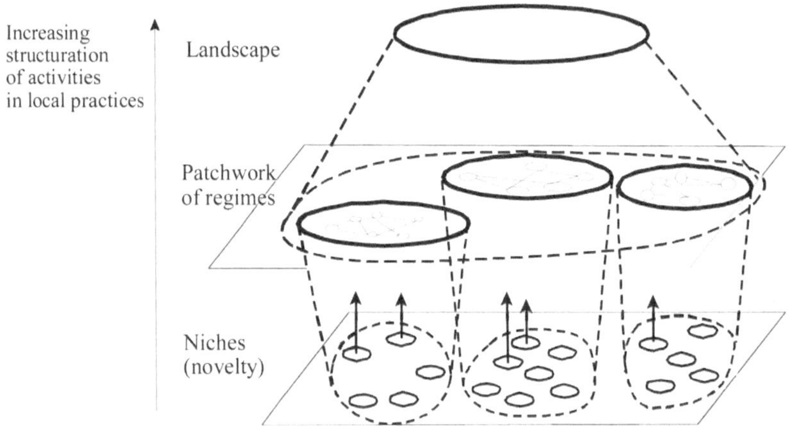

Figure 4. Multiple levels as a nested hierarchy (Geels 2002, 1261).

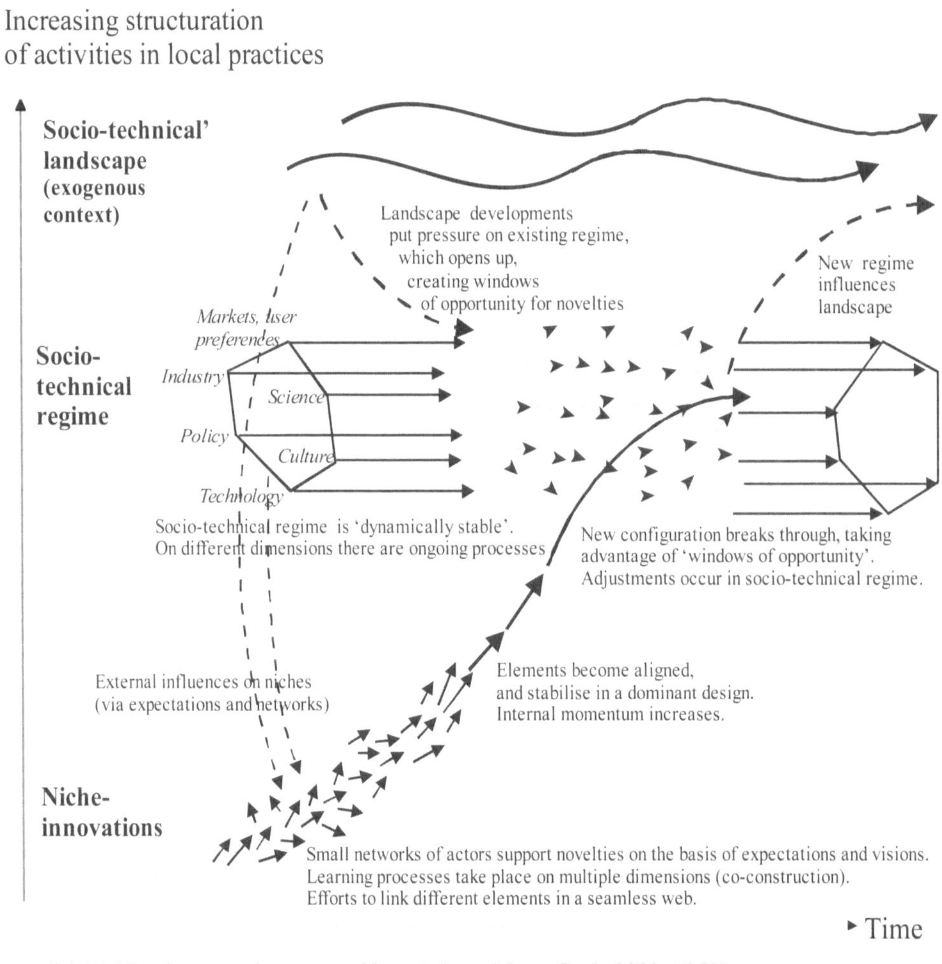

Figure 5. Multi-level perspective on transitions (adapted from Geels 2002, 1263).

The MLP thus corrects the suggestion of the early SNM literature that regime shifts would come about through bottom–up processes of niche expansion (see Figure 1). Instead, alignments of processes at multiple levels are now emphasised. Niche innovations are still important, but they can only diffuse more widely if they link up with ongoing processes at regime and landscape levels. As Shove and Walker (2007, 764) formulated it: 'the key idea is that change takes place through processes of co-evolution and mutual adaptation within and between the layers'.

Recent work distinguishes various types of co-evolution. Raven (2006) found that niche innovations may be adopted from the start within the regime to solve certain problems. Subsequently, further learning processes may lead to more substantial reconfigurations of the regime. Niche innovations thus need not always compete with and substitute for the prevailing regime, as was assumed in earlier SNM work. They may also be incorporated and transform the regime from within. Raven analysed how biomass was incorporated in the electricity production regime through co-firing with coal. Ongoing learning processes and stricter regulatory pressure subsequently triggered further reconfigurations in the electricity regime. Another co-evolution pattern, translation from niche experiences to the regime, was found by Smith (2007) in a study of organic food and green housing in the UK. He found that new practices were initially pioneered by niche actors in relatively secluded spaces (dedicated green activists or architects). Broader regime changes occurred, however, when the niche lessons were translated and picked up by regime actors.

In both co-evolution patterns, the dynamic is less about substitution and more about how niches may branch, pile up, and contribute to changes in the behaviour, practices and routines of existing regime actors. This more differentiated view of niche–regime interaction is fruitful terrain for further research. It also shows that niches can play different roles. They can grow and become a new regime that eventually replaces the old one, but they can also be incorporated into existing regimes. This conclusion led to various systematic attempts to distinguish various types of niches and different transition pathways (Berkhout, Smith, and Stirling 2004; Geels and Schot 2007). Because this is not the right place to elaborate on this work, we only conclude that a next step would be to incorporate the idea of transition pathways more explicitly into SNM work (which may also lead to differentiated policy suggestions).

While the MLP helped to further develop the SNM perspective, SNM research also contributes to the MLP. Most of the MLP studies focus on one regime, but Raven found that niche developments may be influenced by multiple regimes (Raven 2005, 2007; Raven and Verbong 2007).[16] This influence can be beneficial when a niche innovation becomes linked as solution to multiple regimes, but it can also create new problems and uncertainties about regulations, definitions, technical linkages, and responsibilities. The burning of waste in electric utilities, for instance, involved struggles over emission regulations (regarding mercury and other heavy metals), which were stricter for the waste regime than for the electricity regime. There were also struggles over the question of whether organic waste should be defined as biomass (which would make it eligible to renewable energy subsidies) or as waste. Another way to broaden the analytical scope is to study interactions between multiple niche innovations. Sandén and Jonasson (2005), for instance, studied interactions between multiple types of alternative transport fuels. This broader attention to multiple regimes and multiple niches may also lead to modifications in the existing pathway typologies, which are limited to transitions from one regime to another.

5. SNM and its policy implications

The research discussed shows that, contrary to what SNM approaches would favour, many experiments are organised to push for a certain technology and neglect the necessary co-evolutionary dynamics. Furthermore, experiments are often isolated local projects that are not connected to a broader strategy to develop a (global) niche. An important policy question is: how can this structural technology push bias be overcome?[17] This is not an easy question, since the bias is deeply embedded in the modernist way of managing the introduction of technology in society. Ultimately, it would require not only a change in the specific practice of organising experiments, but also broader institutional and cultural changes, particularly in the distribution of responsibilities and the organisation of relations between state, market, civil society and science and technology. This article is not the place to discuss this issue at length (see Schot 2003). We only point to the overlap with Grin's diagnosis, which calls for a new reflexive governance model that appreciates the profound changes that are occurring in the relations between these areas, and conclude with him that to evaluate the policy relevance of SNM and TM, the question needs to be asked to what extent and in what ways they would benefit from, or be hindered by, these profound changes (see Grin 2006). In the remainder of the article, we restrict ourselves to some comments about the nature and limitations of the policy advice generated within SNM research.

SNM was developed to find ways of coping with the policy challenge of nurturing sustainable innovation journeys and transitions. Building on findings of the last 10 years, we conclude that hypotheses about the importance of identified niche internal assumptions are sustained when outcomes of experiments are evaluated *ex-post*. Building on these findings, SNM research has generated a lot of policy advice aimed at creating appropriate processes of network development, learning and visioning. This advice often focuses on generating more appreciation and reflexivity about the ongoing dynamics. It does not result in clear-cut recipes, but helps identify a number of dilemmas. We list of number of them in Table 1. An important contribution of SNM research may thus consist in helping policy makers build competences in recognising and dealing with these policy dilemmas. For this reason we support the initiative of the Dutch Competence Centre for Transitions, in collaboration with the Knowledge Network on Systems Innovations and Transitions, to develop a so-called 'competence kit' on experimentation, which is to be used in real life (Dutch) transition practices (see Raven et al. 2007).[18] It remains to be seen, however, if such instruments actually work in practice and have the intended effects. So far, SNM has been used primarily for *ex-post* evaluations of case studies. It has not been applied prescriptively in ongoing processes.

The research discussed indicates that SNM is not a silver bullet solution that will bring about transitions towards sustainable development, if only because experimenting will not be sufficient. SNM should be seen as a useful addition to existing policy instruments that have neglected the value of experiments. Other more traditional instruments for inducing sustainable innovation, such as market incentives, various forms of regulation, and technology forcing, also have to play a role. Elzen, Hoogma, and Schot (1996) have formulated some initial ideas about the relative influence of different policy strategies on niche-internal development (see also Van der Laak, Raven, and Verbong 2007).

We would like to add a final comment on the position of researchers in this type of action oriented research. SNM suggests that researchers can act as mobilisers, advisors, mappers of change dynamics, and change agents in the name of sustainable development. While SNM recognises that different definitions of sustainable development exist, it is based on the assumption that sustainable development captures enough common ground to act upon. In reaction to this active involvement of SNM (and TM scholars), Shove and Walker have critically asked:

> What are the everyday politics of such an enterprise? When and how are the goals of transition management subject to critical scrutiny, and by whom? Equally important, who wins and who loses out as transition are guided in one direction but not another? (Shove and Walker 2007, 765)

Table 1. Policy dilemmas for niche development.

Expectations, visions	Be flexible, engage in iterative visioning exercises; adjust visions to circumstances and take advantage of windows of opportunity.	Be persistent, stick to the vision, persist when the going gets tough.
Learning	Create variety to facilitate broad learning.	Too much variety dilutes precious resources and prevents accumulation. It also creates uncertainty and may delay choices/commitments (by consumers, policy makers).
Learning	Upscaling through bricolage strategy and stepwise learning. Disadvantages: (1) slow, (2) incremental steps.	Upscaling through breakthrough strategy and big leaps to achieve success rapidly. Disadvantages: (1) danger of failure, (2) mis-alignment with selection environment.
Network	Work with incumbent actors, who have many resources, competence and 'mass'. Try to change their agenda, visions.	For radical innovations, it is better to work with outsiders, who think 'out of the box' and have new ideas. Incumbents have too many vested interests and will try to hinder or encapsulate radical innovations.
Protection	Protection is needed to enable nurturing of niche-innovations.	Do not protect too long and too much. This might lead to limited exposure to selection pressures (and the danger of creating white elephants).
Niche-regime interaction	Wait for 'cracks' in the regime, and then vigorously stimulate niche-innovations. Until such windows of opportunity arise, niches should be nurtured to facilitate stabilisation.	Use niche experiences to influence perceptions of regime actors and actively create cracks in the regime.

These are good questions, because there are politics and power play in SNM processes, the depth of which are easily underplayed. There is no clear solution, however. Independent 'outside positions' do not exist. This is one reason why open-ended learning processes are emphasised in SNM. From this perspective, resistance and conflict is to be expected, and should also be embraced since it may enhance learning processes and allow for the exploration of different futures. Finally, the work on SNM (and TM) originates from a particular assessment of the socio-political situation. While it is clear that investments in R&D and pilot projects for promising sustainable development have increased substantially in the last five years in many sectors (transport, energy, agri-food), at the regime level sustainability is not (yet) the main driver or concern. In the transport regime, for instance, congestion and safety are higher on the agenda than sustainability, even though multi-million R&D programmes are conducted on fuel cells and biofuels. Major car manufacturers also face strong competition, hostile takeovers and rising costs (e.g. pension problems at GM), which receive more attention than sustainability issues. In the electricity regime, liberalisation and privatisation are leading concerns for regime actors. Environmental issues have also appeared on the agenda, but still rank lower than other criteria such as low cost, reliability of supply, and diversification (Verbong and Geels 2007). These regime diagnoses imply that, at the moment, sustainability (still) faces an uphill battle. Although warnings about the political dimension of the SNM and TM research are welcome, the dilemma is that too much reflexivity may lead to paralysis. Political actors who try to deal with the challenge of sustainable development are in need of ideas and approaches that provide handles for addressing the required fundamental changes in the way we live and work. SNM is an answer to this need.

6. Conclusions

We have discussed empirical findings and conceptual elaborations of the last 10 years in SNM research. This discussion has showed that concepts and ideas evolved over time. The internal analysis of crucial niche processes (expectation dynamics, learning, network building) was increasingly complemented with attention to external processes. Particularly, the multi-level perspective proved useful for contextualising SNM. This led to modifications in claims about the breakthrough of sustainable innovation journeys. Niches and experiments are to be perceived as crucial for bringing about regime shifts, but not as the sole forces in doing so. Linkages with ongoing processes at broader regime and landscape levels are important. With regard to interactions between niches and regimes, initial work only focused on technological substitution, but recent work also distinguishes other types of interaction (e.g. incorporation, translation) and more differentiated transition pathways (e.g. reconfiguration, transformation). In sum, the understanding of niche dynamics has become increasingly complex compared to simple technical substitution ideas in the mid 1990s.

Although substantial insights have been gained, the SNM approach is still an unfinished research programme. Therefore, we would like to end by listing a number of promising research issues identified in this article:

- The role of visions in the process of niche formation, including their influence on the building of positive feedbacks between changes at the niche, regime and landscape level;
- The nature and source of protection of niches that is conducive to its further development, as well as the management of selective exposure;
- The mechanisms that make sequences of projects gel into niche development;
- The way interactions among multiple niches and multiple regimes influence niche development patterns;
- Action research of prescriptive applications of SNM;
- A systematic comparison SNM, TM, and other policy measures.

In seems fair to conclude that SNM has already become a niche in policy studies about sustainable innovation journeys, but further work and implementation is needed to enhance the prospect of sustainability transitions and the larger issue of transforming the modernist approach of regulating impacts and correcting market failures through subsidies (see also Nill and Kemp 2008, forthcoming).

Acknowledgement

We have had the great fortune to discuss this article with a wide range of scholars, many heavily involved themselves in SNM and transition research and therefore cited in the text. We thank them for their constructive criticism. We gratefully acknowledge the support of the Dutch Knowledge Network on System Innovation and Transitions towards Sustainable Development (KSI).

Notes

1. Van de Belt and Rip (1984) did not address sustainable development and regime change, but nurturing, survival and robustness. They introduced SNM as a combination of protection against too harsh selection at an early stage and subsequent controlled exposure to selection pressures.
2. For a nice overview, see Williams and Edge (1996).
3. For this historical analysis, see Schot (2003).
4. See Rip and Kemp (1998) for the idea of modulating ongoing interactions; for the notion of steering from within, see Rip (2006). Kemp developed the idea of modulating ongoing dynamics into an evolutionary policy approach; see Nill and Kemp (2008, forthcoming).
5. When Rip (1992, 1995) introduced the niche concept, he used the example of the R&D programme on the Stirling engine inside the Philips Company. His basic point was not that niches are internal to firms, however, but that radical novelties initially have low legitimacy and require protection and nurturing to survive (e.g. dedicated product development programmes). See also Van den Belt and Rip (1987). Schot, Hoogma, and Elzen (1994), expanded the idea and introduced the distinction between technological and market niche, and proposed to use the notion of technological niche to refer to societal experiments with new technologies outside the laboratory in a user context.
6. On this point, see Verheul and Vergragt (1995).
7. On the role of expectations, see Brown and Michael (2003); see also Van Lente (1993).
8. For an overview of various conceptualisations of niches in radical change see Schot and Geels (2007).
9. This division in three processes has been much discussed, leading to various proposals. For example, Hoogma et al. (2002) propose to distinguish two main processes; learning processes and institutional embedding, with further subdivisions (pp. 28–29).
10. In 1998, the European Union funded an SNM research project within the 'Environment and Climate' RTD programme. Through this project, scholars in several countries worked on SNM. They investigated fourteen innovative transport projects in different European cities (ranging from electric vehicles to car sharing schemes). This collaborative project resulted in a work book for practitioners on how to do SNM (Weber et al. 1999), and an academic book (Hoogma et al. 2002).
11. The following articles explicitly applied the SNM perspective: Brown et al. (2004), Ieromonachou, Potter, and Enoch (2004), Truffer, Metzner, and Hoogma (2004), Kivisaari, Lovio, and Väyrynen (2004), Harborne, Hendry, and Brown (2007), Hendry (2007), Van Eijck and Romijn (2008) and Hegger et al. (2007). Two other PhD theses that discuss SNM are Lane (2002) and Adey (2007). Finally SNM is central to Wiskerke and Van der Ploeg (2002).
12. TM partly draws on SNM. Kemp has been the intermediary between the two approaches. Another approach that highlights the role of visions is backcasting (see Quist 2007). For a reflection on SNM and TM, see Loorbach and Van Raak (2007).
13. This is the theme of a thesis under preparation by Ulmanen. For first results see Ulmanen, Raven and Verbong (2007).
14. For results published in English, see Schot (1998) and Van Driel and Schot (2005).
15. For a further discussion on the nature of this macro-level, see Van Driel and Schot (2005) and Geels and Schot (2007).
16. For multi-regime interaction in the multi-level perspective, see Geels (2007).
17. Policies induced by Transition management encounter a similar problem, see Kemp, Rotmans, and Loorbach (2007).
18. Two other books have been produced for practitioners but with less direct involvement of them (Kemp and Van den Bosch 2006; Weber et al. 1999). Insights of the later working book are elaborated in Weber and Dorba (1999). More information on www.transitiepraktijk.nl and www.ksinetwork.nl

References

Adey, S. 2007. A journey without maps. Towards sustainable subsistence agriculture in South Afrika, PhD thesis, Wageningen University, Wageningen.

Astley, W.G. 1985. The two ecologies: population and community perspectives on organizational evolution. *Administrative Science Quarterly* 30, no. 2: 224–41.

Berkhout, F., A. Smith, and A. Stirling. 2004. Socio-technological regimes and transition contexts. In *System innovation and the transition to sustainability: theory, evidence and policy*, eds. B. Elzen, F.W. Geels and K. Green, 48–75. Cheltenham: Edward Elgar.

Bijker, W.E. 1995. *Of bicycles, bakelites and bulbs: towards a theory of sociotechnical change*. Cambridge, MA: The MIT Press.

Bos, B., and J. Grin. 2008. Doing reflexive modernisation in pig husbandry; the hard work of changing the course of a river. *Science, Technology and Human Values* 33, no. 4: 480-507.

Brown, H.S., P.J. Vergragt, K. Green, and L. Berchicci. 2004. Bounded socio-technical experiments (BSTEs): higher order leaning for transitions towards sustainable mobility. In *System innovation and the transition to sustainability: theory, evidence and policy*, eds. B. Elzen, F.W. Geels and K. Green, 48–75. Cheltenham: Edward Elgar.

Brown, N., and M. Michael. 2003. The sociology of expectations: retrospecting prospects and prospecting retrospects. *Technology Analysis & Strategic Management* 15, no. 1: 3–18.

Deuten, J.J. 2003. Cosmopolitanising technology: a study of four emerging technological regimes. PhD thesis, Twente University Press, Enschede.

Dosi, G. 1982. Technological paradigms and technological trajectories: a suggested interpretation of the determinants and directions of technical change. *Research Policy* 11, no. 3: 147–62.

Elzen, B., R. Hoogma, and J. Schot. 1996. Mobiliteit met Toekomst; Naar een vraaggericht technologiebeleid [Mobility with a future. Towards a demand-oriented technology policy]. Report to the Ministry of Traffic and Transport (in Dutch). Adviesdienst Verkeer en Vervoer, Rijkswaterstaat, Rotterdam.

Frenken, K., P.P. Saviotti, and M. Trommetter. 1999. Variety and niche creation in aircraft, helicopters, motorcycles and minicomputers. *Research Policy* 28, no. 5: 469–88.

Geels, F.W. 2002. Technological transitions as evolutionary reconfiguration processes: a multi-level perspective and a case-study. *Research Policy* 31, no. 8/9: 1257–74.

———. 2005. *Technological transitions and system innovations: a co-evolutionary and socio-technical analysis*. Cheltenham: Edward Elgar.

———. 2007. Analysing the breakthrough of rock 'n' roll (1930–1970): multi-regime interaction and reconfiguration in the multi-level perspective. *Technological Forecasting and Social Change* 74, no. 8: 1411–31.

Geels, F.W., and J. Schot. 2007. Typology of sociotechnical transition pathways. *Research Policy* 36, no. 3: 399–417.

Geels, F.W., and J.J. Deuten. 2006. Local and global dynamics in technological development: a socio-cognitive perspective on knowledge flows and lessons from reinforced concrete. *Science and Public Policy* 33, no. 4: 265–75.

Geels, F.W., and R.P.J.M. Raven. 2006. Non-linearity and expectations in niche-development trajectories: ups and downs in Dutch biogas development (1973–2003). *Technology Analysis & Strategic Management* 18, no. 3/4: 375–92.

Grin, J. 2006. Reflexive modernisation as a governance issue, or: designing and shaping re-structuration. In *Reflexive governance for sustainable development*, eds. J.-P. Voss, D. Bauknecht and R. Kemp, 57–81. Cheltenham: Edward Elgar.

Grin, J., and H. van de Graaf. 1996. Implementation as communicative action: an interpretive understanding of the interactions between policy makers and target groups. *Policy Sciences* 29, no. 4: 291–319.

Harborne, P., C. Hendry, and J. Brown. 2007. The development and diffusion of radical technological innovation: the role of bus demonstrations projects in commercializing fuel cell technology. *Technology Analysis and Strategic Management* 19, no. 2: 167–88.

Hegger, D.L.T., J. van Vliet, and B.J.M. van Vliet. 2007. Niche management and its contribution to regime change: the case of innovation in sanitation. *Technology Analysis and Strategic Management* 19, no. 6: 729–46.

Hendry, C., P. Harborne, and J. Brown. 2007. Niche entry as a route to mainstream innovation: learning from the phosphoric acid fuel cell in stationary power. *Technology Analysis & Strategic Management* 19, no. 4: 403–25.

Hommels, A., P. Peters, and W.E. Bijker. 2007. Techno therapy or nurtured niches? Technology studies and the evaluation of radical innovations. *Research Policy* 36, no. 7: 1088–99.

Hoogma, R. 2002. Exploiting technological niches: strategies for experimental introduction of electric vehicles. PhD thesis, Twente University, Enschede.

Hoogma, R., R. Kemp, J. Schot, and B. Truffer. 2002. *Experimenting for sustainable transport: the approach of strategic niche management*. London: Spon Press.

Ieromonachou, P., S. Potter, and M. Enoch. 2004. Adapting strategic niche management for evaluating radical transport policies: the case of the Durham road access charging scheme, *International Journal of Transport Management* 2, no. 2: 75–87.

Kemp, R., and S. van den Bosch. 2006. *Transitie-experimenten: praktijkexperimenten met de potentie om bij te dragen aan transitie*. Rotterdam: Kenniscentrum voor duurzame systeeminnovaties en transities.

Kemp, R., J. Rotmans, and D. Loorbach. 2007. Assessing the Dutch energy transition policy: how does it deal with dilemmas of managing transitions. *Journal of Environmental Policy and Planning* 9, no. 3/4: 315–31.

Kemp, R., J. Schot, and R. Hoogma. 1998. Regime shifts to sustainability through processes of niche formation: the approach of strategic niche management. *Technology Analysis and Strategic Management* 10, no. 2: 175–96.

Kivisaari, S., R. Lovio, and E. Väyrynen. 2004. Managing experiments for transition: examples of societal embedding in energy and health care sectors. In *System innovation and the transition to sustainability: theory, evidence and policy*, eds. B. Elzen, F.W. Geels and K. Green, 223–50. Cheltenham: Edward Elgar.
Lane, B. 2002. Implementation strategies for fuel-cell powered road transport systems in the United Kingdom. PhD thesis, Open University, Milton Keynes.
Levinthal, D.A. 1998. The slow pace of rapid technological change: gradualism and punctuation in technological change. *Industrial and Corporate Change* 7, no. 2: 217–47.
Loorbach, D. 2007. Transition management: new mode of governance for sustainable development, PhD thesis, Erasmus University Rotterdam.
Loorbach, D., and R. van Raak. 2007. Strategic niche management and transition management: different but complementary approaches. Internal manuscript DRIFT, Erasmus University, Rotterdam.
Lynn, G.S., J.G. Morone, and A.S. Paulson. 1996. Marketing and discontinuous innovation: the probe and learn process. *California Management Review* 38, no. 3: 8–37.
Mokyr, J. 1990. *The lever of riches: technological creativity and economic progress*. New York: Oxford University Press.
Nelson, R.R., and S.G. Winter. 1982. *An evolutionary theory of economic change*. Cambridge, MA: Harvard University Press.
Nill, J., and R. Kemp. 2008. Evolutionary approaches for sustainable innovation policies: from niche to paradigm. *Research Policy*, forthcoming.
Quist, J. 2007. Backcasting for a sustainable future. The impact after 10 years. PhD thesis, Technical University Delft.
Raven, R., S. van den Bosch, G. Fonk, J. Andringa, and R. Weterings. 2007. *Competentiekit experimenteren*. Utrecht: Competentiecentrum Transities.
Raven, R.P.J.M. 2005. Strategic niche management for biomass. PhD thesis, Eindhoven University of Technology, The Netherlands.
———. 2006. Towards alternative trajectories? Reconfigurations in the Dutch electricity regime. *Research Policy* 35, no. 4: 581–95.
———. 2007. Co-evolution of waste and electricity regimes: multi-regime dynamics in the Netherlands (1969–2003). *Energy Policy* 35, no. 4: 2197–2008.
Raven, R.P.J.M., and G.P.J. Verbong. 2007. Multi-regime interactions in the Dutch energy sector. The case of combined heat and power in the Netherlands 1970–2000. *Technology Analysis and Strategic Management* 19, no. 4: 491–507.
Rip, A. 1992. A quasi-evolutionary model of technological development and a cognitive approach to technology policy. *Rivista di Studi Epistemologici e Sociali Sulla Scienza e la Tecnologia* 2: 69–103.
———. 1995. Introduction of new technology: making use of recent insights from sociology and economics of technology. *Technology Analysis & Strategic Management* 7, no. 4: 417–31.
———. 2006. A co-evolutionary approach to reflexive governance – and its ironies. In *Reflexive governance for sustainable development*, eds. J.-P. Voss, D. Bauknecht and R. Kemp, 82–100. Cheltenham: Edward Elgar.
Rip, A., and R. Kemp. 1998. Technological change. In *Human choice and climate change*, eds. S. Rayner and E.L. Malone, vol. 2, 327–399. Columbus, OH: Battelle Press.
Rosenberg, N. 1976. *Perspectives on technology*. Cambridge: Cambridge University Press.
Rotmans, J., R. Kemp, and M. van Asselt. 2001. More evolution than revolution: transition management in public policy. *Foresight* 3, no. 1: 15–31.
Sandén, B., and K. Jonasson. 2005. Competition and co-evolution among contenders: the development of alternative transport fuels in Sweden 1974–2004. Paper presented at 4th European Meeting on Applied Evolutionary Economics, Utrecht, The Netherlands, 19–21 May 2005.
Saviotti, P.P. 1996. *Technological evolution, variety and the economy*. Cheltenham: Edward Elgar.
Schot, J. 1992. The policy relevance of the quasi-evolutionary model: the case of stimulating clean technologies. In *Technological change and company strategies: economic and sociological perspectives*, eds. R. Coombs, P. Saviotti and V. Walsh, 185–200. London: Academic Press.
———. 1998. The usefulness of evolutionary models for explaining innovation. The case of the Netherlands in the nineteenth century. *History and Technology* 14: 173–200.
———. 2003. The contested rise of a modernist technology politics. In *Technology and modernity*, eds. Th.J. Misa, P. Brey and A. Rip, 257–78. Cambridge: MIT Press.
Schot, J., and F.W. Geels. 2007. Niches in evolutionary theories of technical change. *Journal of Evolutionary Economics* 17, no. 5: 605–22.
Schot, J., R. Hoogma, and B. Elzen. 1994. Strategies for shifting technological systems. The case of the automobile system. *Futures* 26, no. 10: 1060–76.

Shove, E., and G. Walker, G. 2007. CAUTION! Transitions ahead: politics, practice and sustainable transition management. *Environment and Planning A* 39, no. 4: 763–70.
Smith, A. 2007. Translating sustainabilities between green niches and socio-technical regimes. *Technology Analysis and Strategic Management* 19, no. 4: 403–25.
Truffer, B., A. Metzner, and R. Hoogma. 2004. The coupling of viewing and doing: strategic niche management and the electrification of individual transport. *Greener Management International* 37: 111–24.
Ulmanen, J., R.P.J.M. Raven, and G.P.J. Verbong. 2007. Creating legitimacy for sustainable technology development. The case of Dutch biofuels for the transport sector. Paper for the 15th European Biomass Conference and Exhibition, 7–11 May 2007, Berlin.
Van de Belt, H., and A. Rip. 1984. Technologie-ontwikkeling: het Nelson–Winter/Dosi model. A report to the Government Office of Science Policy. LISBON, University of Leiden.
———. 1987. The Nelson–Winter–Dosi model and synthetic dye chemistry. In *The social construction of technological systems: new directions in the sociology and history of technology*, eds. W.E. Bijker, T.P. Hughes and T. Pinch, 135–58. Cambridge, MA: The MIT Press.
Van den Bosch, S., and M. Taanman. 1996. How innovation impacts society. Patterns and mechanisms through which innovation projects contribute to transitions. Paper presented at Innovation Pressure Conference, 15–7 March 1996, Tampere, Finland.
Van der Laak, W., R.P.J.M. Raven, and G.P.J. Verbong. 2007. Strategic niche management for biofuels. Analysing past experiment for developing new biofuels policy. *Energy Policy* 35, no. 6: 3213–25.
Van Driel, H., and J. Schot. 2005. Radical innovation as a multi-level process: introducing floating grain elevators in the port of Rotterdam. *Technology and Culture* 46, no. 1: 51–76.
Van Eijck, J., and H. Romijn. 2008. Prospects for Jatropha Biofuels in Tanzania: an analysis with strategic niche management. *Energy Policy* 36, no. 1: 311–25.
Van Lente, H. 1993. Promising technology: the dynamics of expectations in technological development. PhD thesis, Eburon, Twente University, Delft.
Van Mierlo, B. 2002. Kiem van Maatschappelijke Verandering: Verspreiding van Zonnecelsystemen in de Woningbouw met behulp van Pilot Projecten. PhD thesis, University of Amsterdam.
Verbong, G.P.J., and F.W. Geels. 2007. The ongoing energy transition: lessons from a socio-technical, multi-level analysis of the Dutch electricity system (1960–2004). *Energy Policy* 35, no. 2: 1025–37.
Verheul, H., and P.J. Vergragt. 1995. Social experiments in the development of environmental technology: a bottom–up perspective. *Technology Analysis and Strategic Management* 7: 315–26.
Voss, J.-P., D. Bauknecht, and R. Kemp, eds. 2006. *Reflexive governance for sustainable development*, 82–100. Cheltenham: Edward Elgar.
Weber, M., R. Hoogma, B. Lane, and J. Schot. 1999. *Experimenting with sustainable transport innovations: a workbook for strategic niche management*. Seville/Enschede: Universiteit Twente.
Weber, W., and A. Dorba. 1999. Strategic niche management: a tool for the market introduction of new transport concepts and technologies. The IPTS Report 31, Sevilla, pp. 20–8.
Williams, R., and D. Edge. 1996. The social shaping of technology. *Research Policy* 25, no. 6: 865–99.
Windrum, P., and C. Birchenhall. 1998. Is product life-cycle theory a special case? Dominant designs and the emergence of market niches through co-evolutionary learning. *Structural Change and Economic Dynamics* 9, no. 1: 109–34.
Wiskerke, J.S.C., and J.D. van der Ploeg. 2002. *Seeds of transition: essays on novelty production niches and regimes in agriculture*. Assen: Van Gorcum.

Multi-niche analysis of dynamics and policies in Dutch renewable energy innovation journeys (1970–2006): hype-cycles, closed networks and technology-focused learning

Geert Verbong, Frank W. Geels and Rob Raven

Eindhoven University of Technology, The Netherlands

This article analyses long-term innovation policies and development trajectories of four renewable energy technologies: wind energy, biomass, fuel cells and hydrogen, and photovoltaics. These trajectories and policies are characterised by many costly failures, setbacks, hype-disappointment cycles, tensions, and struggles. Although setbacks and non-linearities are a normal part of innovation journeys, a comparative analysis of four cases shows the recurrence of particular problems. Using Strategic Niche Management as analytical approach, we conclude that major problems exist with regard to learning processes (too much technology-push, focused on R&D), social networks (supply side oriented, narrow, closed) and expectations (hype-disappointment cycles, limited competence to assess promises).

1. Introduction

Although there is much talk in the Netherlands about energy transitions, implementation of renewable energy is low compared to other European countries (Figure 1).

The low Dutch share of renewable energy is not for lack of effort. Since the 1970s, there have been many policy activities, investments and projects to develop and implement renewable energy options. The Environmental Policy Plan (1989) and the Third Energy White Paper (1995) paid much attention to environmental problems and climate change and formulated high ambitions (e.g. 10% renewable energy in 2020). But there has been a large gap between policy goals and the implementation of substantive policy measures to reach those goals (Verbong and Geels 2007). There have been many costly failures, setbacks, hype–disappointment cycles, tensions, and struggles in innovation journeys with renewable energy innovations. This is

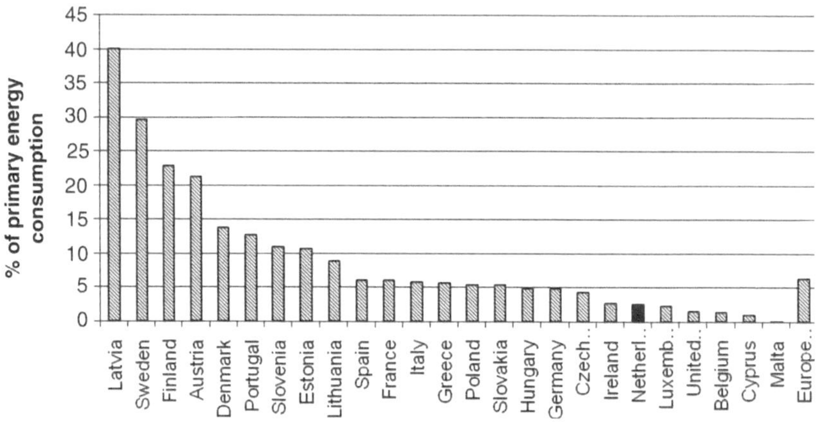

Figure 1. Share of renewable energies in primary energy consumption in European Union countries 2005 (EuroObserv'ER).

not necessarily remarkable, because innovation journeys with new technologies are always non-linear and unpredictable (Van de Ven et al. 1999; Geels and Raven 2006). Innovation journeys without some failures and setbacks do not exist. Renewable energy innovation in the Netherlands, however, seems to suffer from *recurring* failures and problems. Hence, the article aims: (1) to provide an empirical analysis of renewable energy innovation journeys, with particular attention for recurring patterns and mechanisms, (2) derive policy lessons from the analysis.

The analysis uses strategic niche management (SNM) as conceptual framework. SNM is based on a combination of two theories of technological change: social constructivism and evolutionary economics (see also Schot and Geels in this special issue). The social construction of technology (SCOT) approach analyses technological change as the outcome of interactions between various social groups (e.g. researchers, firms, policy makers, users, special-interest groups) (Bijker, Hughes, and Pinch 1987). SCOT particularly highlights socio-cognitive dimensions such as interpretations and beliefs. Because new technologies are characterised by 'interpretative flexibility', social groups often have different ideas and views. These ideas and interpretations influence the actions of actors and social groups with regard to investments, subsidies, pilot projects, contestations, and struggles. Social coalitions and cognitive interpretations may eventually stabilise as a result of learning processes and network building. SNM builds on SCOT, seeing technological change as socially enacted process, in which learning (changing interpretations) and network building are two crucial processes. Building on the sociology of expectations (Van Lente 1993; Brown and Michael 2003), SNM adds a third process: the articulation of expectations. These future-oriented visions not only guide actor's activities, but can also be used to interest and enrol other actors, thus broadening the social networks.

From evolutionary economics, SNM borrows the idea that new technologies require protection from mainstream market selection. This idea comes from biological evolution theory, which argues that new species emerge in geographical niches that are separated from a parent population (Mayr 1963). For technological change, such niches are socially constructed, consisting of social networks and actors that are willing to invest time and resources in a fledgling technology, because they believe in its potential (Schot and Geels 2007). In the case of renewable energy innovations, for instance, policy makers have provided such protection through subsidies and regulatory adaptations. SNM has further evolutionary connotations because of its attention for the alignment of variations (new technologies that often arise from R&D) and selection environment. SNM particularly focuses on bridging the 'valley of death' between R&D and market introduction. SNM-scholars have shown that concrete and real-life experimental projects (R&D, pilot, and demonstration projects) play an important role in that respect (Kemp, Schot, and Hoogma 1998; Hoogma et al. 2002; Raven 2005; Geels and Raven 2006). These projects bring together actors from the variation environment (researchers, firms, technology developers) and selection environment (users, policy makers, special interest groups), facilitate network building, stimulate learning processes and produce outcomes that may lead to adjustments in expectations.

The combination of SCOT and evolutionary economics allows SNM to open up the black box of technology. While evolutionary economics tends to operationalise 'learning' as R&D investments, patents and information accumulation, SNM additionally looks at the *content* of learning and the changes over time in perceptions, ideas and cognitive frames. With regard to networks, SNM not only looks at knowledge flows and exchange (of products, money and information), but also at *social* dimensions such as coalitions, trust, conflicts, and tensions.

Using SNM as conceptual framework, Section 2 provides 30-year descriptions of four niche-innovation trajectories: wind turbines, biomass, fuel cells and hydrogen, and photovoltaic (PV). These descriptions focus on the three main SNM-concepts: (1) social networks dynamics related to innovations; (2) expectation dynamics (visions, claims and counterclaims), and (3) learning processes. Section 3 provides a more systematic analysis of recurring patterns and mechanisms across the four cases. Section 3 also formulates policy conclusions.

Given the social enactment characteristics of the theoretical framework, the analysis belongs to the 'process theory' tradition in social science. Recent articles and special issues distinguish process theory from variance theory (Petttigrew 1997; Van de Ven and Poole 2005).[1] *Variance theory* explains (change in) outcomes as the product of independent variables (causal factors) acting on dependent variables. *Process theory* focuses on *events* rather than causal factors (Abbott 1992). Events are enacted by actors who make decisions, undertake actions and react to each other. Taking the notion of path dependence seriously, process theories explain outcomes as the result of temporal *sequences of events* and the *timing* and *conjunctures* of event-chains.

Process theory assumes that the world is made up of entities (people, organisations, technologies) that participate in events ('enactment') and may change their interpretations, preferences and networks. Because of this ontology, process theories use particular methodologies. *Variance theories*, which aim to estimate the relative importance of causal factors, usually collect quantitative data (facts, figures) that can be processed with statistical methods. *Process theories*, on the other hand, tend to adopt qualitative, interpretive methods (e.g. case studies), because they are rich in context and allow 'process tracing' of event chains and social interactions. Process theorists often use narrative methods, because they can capture fuzzy and unfolding processes (Abbott 1992; Griffin 1993). Some narratives, however, only describe 'one damn thing after another'. To turn case *histories* into case *studies*, narrative plots should be used to provide more analytical focus (Pentland 1999).

Pettigrew (1990) suggests further analytical steps. Instead of case histories, he advises scholars to write *analytic chronologies*. While an analytical chronology lays out the narrative, 'there is an explicit attempt to interpret and explain' the process (p. 280). The aim is not to tell the story 'as it really was' (an unrealistic ambition, partly, but not only, because of space constraints), but to use narrative plots to provide more selective analytical focus. Section 2 uses the three main SNM processes in that respect. Pettigrew (1990, 280) also suggests a further step towards *interpretative theoretical cases* that make 'both a more explicit attempt to interpret the narrative but also to link emerging conceptual and theoretical ideas inductively derived from the case to stronger analytical themes'. In that respect, Section 3 identifies more general analytical patterns and mechanism.

With regard to data, we use secondary sources, refer to academic publications in the literature, and draw from our own work, which is based on archival research (institutional and private archives), interviews, journals, reports, books, newspapers and internet sites. Elaborate descriptions of this research can be found elsewhere, e.g. studies of the electricity system (Verbong and Geels 2007), renewable energy (Verbong et al. 2001), co-firing of biomass and coal (Raven 2006), and biomass digestion (Geels and Raven 2006).

2. Description of niche-innovation trajectories

2.1. *Wind energy*

Wind energy was regarded as a promising renewable energy option for the Netherlands. Despite continuous efforts from the 1970s onwards, wind energy still is a minor niche, although Dutch offshore potential is considerable. Moreover, it faces several serious problems. The difficult development trajectory can be explained by a combination of a technology push approach, tensions in the social network and limited learning, resulting in a tarnished symbolic image.

Expectations dynamics

Following the first energy crisis in 1973, policy interest in wind energy rapidly increased. The first national energy research program on wind energy started in 1976. The official goal was to investigate the potential of wind energy, but the general vision was that the Netherlands (and Dutch industry) should become a global leader in this field. References to the glorious past of Dutch windmills served to legitimate a national research program. The outcome of this first program was confirmation of the view that wind energy was a promising option and recommendations for more research. In the early 1980s, a national goal for wind energy was agreed upon: 1000 MW in 2000. Although it became obvious during the 1990s that this target would not be met, it remained the official target (Verbong et al. 2001).

Because actors from the existing electricity regime dominated the social networks (see below), certain ideas guided the visions on wind energy. There was consensus that large-scale turbines, preferably clustered in wind parks, were the best way forward. However, the grass roots movement contested this large-scale vision. It succeeded in creating a decentralised line within the research program, focusing on small autonomous turbines. Learning processes in the late 1980s and early 1990s ran into problems for both designs (see below). Experiences in Denmark showed that gradual upscaling of small turbines was a better innovation journey (Kamp, Smits, and Andriesse 2004). Although Dutch actors also embraced this idea in the mid 1990s, it was too late to save the Dutch wind turbine industry.

In the late 1990s, concerns about climate change stimulated wind energy, which was (still) regarded as one of most important Dutch renewable energy options, but the implementation of wind energy encountered many problems such as local opposition and a weakening of its symbolic image. Public opinion expressed more concerns about landscape disturbance (wind turbines obstructing wide 'open sky' views) and 'bird shredding'. Grid operators increasingly saw wind turbines as unreliable and a danger to the stability of the electricity system (because they provide electricity on an intermittent basis) and politicians came to perceive wind turbines as a waste of tax money and a not very cost-effective way of reducing greenhouse gases.

Social network

The social network in the 1970s consisted mainly of representatives of the incumbent electricity regime: research institutes, large companies and electric utilities. These actors thought that only large turbines were a feasible option. Hence, R&D focused on large MW-size turbines, based on design knowledge from aeronautics. From the start the utilities played an ambiguous role (Verbong 1999). They were initially sceptical about wind energy's potential but sympathetic to (subsidised) initiatives.

The first tensions in the network were played out in the early 1980s. The utilities argued that they should have control over the first wind park, because its functional characteristics were similar to power plants. They succeeded in wrestling the park out of the national program framework, creating frustrations amongst other actors. Relationships between the utilities and technology suppliers further deteriorated, when the park failed because of unreliable wind turbines. The company that supplied the turbines almost went bankrupt. Another industrial company quit participation, complaining about the turbine selection procedure and arrogant behaviour of the electricity sector. The result of these negative experiences was that the attitude of the electricity sector and large industrial companies towards wind energy became more hostile (Verbong 1999).

Despite the disappointments and tensions in the 1980s, the Dutch position around 1990 was still good in an international perspective (Bergek and Jacobsson 2003): (1) there were three middle-sized turbine producing companies, supported by a few producers of rotor blades and other components; (2) the knowledge infrastructure was well developed, especially at Delft university and ECN (the Energy Research Centre of the Netherlands); (3) the utilities formed a foundation for the large-scale implementation of wind turbines, with financial support from the government, and (4) the government reached an agreement on wind turbine sites with seven Provinces. Ten years later, however, ambitions were shattered and the Dutch wind industry was overrun by Danish turbine producers. Although the Dutch companies switched to the Danish design, they could not keep up with the international companies who benefitted from first mover advantages and scale economies (Kamp, Smits, and Andriesse 2004).

The 1989 Electricity Law enforced a separation of production of electricity. While utilities remained responsible for production, the law introduced Energy Distribution Companies (EDCs) as a new actor, who were responsible for electricity distribution and consumer markets. This new actor provided new dynamics because EDCs stimulated wind energy (and other renewable sources) (Verbong and Geels 2007). Although EDCs put most of their efforts and resources in cogeneration of heat and power (CHP), they also stimulated the 'green electricity' market through advertising and image-building campaigns and constructed several wind energy parks (Hofman 2005). Nevertheless, the government did not achieve its official target. Only 450 MW of wind turbines were installed in 2000 (instead of 1000 MW).

The implementation problems were related to increasing tensions between the national policy level, where ambitious targets were set, and the local policy level, which was responsible for implementation, siting and permit procedures. The national level, which focused exclusively on the contributions of wind energy to climate change policies, often neglected possible benefits, sensitivities and opportunities on the local level. Because policy makers focused primarily on large scale applications and regime actors, they neglected broad-based institutional capability building (Breukers 2007).

Learning processes

In the 1970s and 1980s, learning focused mainly on technological aspects. The development of large breakthrough designs proved to be difficult (Kamp, Smits, and Andriesse 2004). Outcomes of learning processes were disappointing, because most large turbines broke down or did not function. Danish firms, in contrast, conquered the market with a bottom–up approach: they started with small-size turbines, and subsequently upscaled them (Van Est 1999). The innovative two-bladed design, favoured by the Dutch (and also Americans and Germans) lost out to the more conservative three-bladed Danish design (Kamp et al. 2004).

Failures also occurred in decentral designs, because the assumption that wind energy systems could function reliably without grid back-up, proved to be erroneous. These learning processes hardened the vision that only grid-connected wind parks had a future. Hence, the relationship with the existing electricity regime was seen as crucial for further development of wind energy (Verbong 2001).

Another issue was wind energy's impact on the electricity system. Because of wind fluctuations and unpredictability, the electricity sector estimated that only 650 MW of wind turbine capacity could be connected without endangering the system. This claim was contested by ECN, which made higher estimates (2500 MW) (Verbong 2001). The underlying issue was (and is) that wind turbines require the presence of spare capacity which can provide backup during windless periods. Normally, utilities paid electricity suppliers for the amount of electricity delivered to the grid, topped up with a 'capacity fee' based on savings in installed capacity. But for wind turbines, the electricity sector did not want to pay such a fee, arguing that they did not really replace existing production units. Wind energy promoters fiercely contested this principle, because low feedback tariffs increased the price gap between wind energy and regular electricity. The principle thus formed an important barrier for local initiatives by co-operatives and private wind associations. Despite these protests, utilities refused to pay 'capacity fees'. Another financial barrier for wind energy suppliers came from additional costs that utilities charged for connecting wind turbines to the grid. Both problems soured relationships in the social network behind wind energy.

In 1996, some conditions improved because of the introduction of an energy tax (REB). Because renewable energy was exempted from this tax, the price gap between 'green' electricity and normal 'grey' electricity was reduced. The tax exemption was supplemented by 2 cents/kWh for production support (Van Rooijen and Van Wees 2006). These support measures enabled wind energy to overcome the two previously mentioned economic barriers (the unpaid capacity fee and grid connection costs). Now 'green' electricity was about the same price as 'grey' electricity, which stimulated households' demand: the number of green electricity consumers rapidly increased from 16,000 in 1996 to 1.4 million in 2002 (Van Rooijen and Van Wees 2006).

From 1996 onwards, contributions from wind energy grew steadily, but the expansion also led to new problems, especially in the process of societal embedding. Local environmental and local interest groups increasingly opposed new wind energy projects, complaining that wind turbines disturbed the natural landscape, acted as 'bird shredders', and were noisy, ugly objects. These protests were the main reasons that wind energy targets for 2000 were not met. After 2000, offshore wind parks were seen as a promising solution, because they avoided on-land problems, but permit-procedures, negotiations with nature conservationists and uncertainty over financial support schemes delayed these projects.

This uncertainty was due to new policy changes in 2003. The underlying problem was that the rapid increase in consumer demand for 'green electricity' could not be met with national production alone. Hence, imports of renewable electricity increased rapidly from 1.5 GWh in

2000 to 10.5 GWh in 2002 (www.cbs.nl). In 2002, only 26% of 'green electricity' was produced domestically (Van Rooijen and Van Wees 2006). The unintended effect of renewable energy subsidies was that Dutch tax money flowed to international suppliers. Hence, a new government stopped the demand-oriented REB exemptions in 2003, replacing it with the supply-oriented and technology specific MEP-regulation (Environmental Quality Electricity generation). The MEP provided a fixed feed-in tariff to renewable electricity producers plus an additional ecotax exemption. The new MEP-scheme was initially set for 10 years, but within 2 years the minister of Economic Affairs announced a major downscaling, because the number of proposals for wind parks was much larger than expected, something that would imply major increases of the MEP budget. The minister thus excluded new offshore wind and large-scale biomass projects from MEP scheme. Only two offshore wind parks acquired financial support. In another twist to this story, the government abruptly announced the end of the MEP-scheme for all new projects, including small-scale projects, in August 2006. The government argued that the Netherlands would reach the EU goal (9% renewable electricity in 2010) with existing projects. Continuation of the MEP-regulation was thus seen as too costly and not really necessary.

These frequent changes in regulations and subsidy schemes and the refusal to support the industry over an extended period of time (as the Germans did) have given the national government an image of unreliability. Moreover, a persistent problem is the neglect of societal embedding of wind energy, with policy focusing primarily on the technical side of innovation (Breukers 2007).

2.2. *Biomass*

Biomass exists in several forms (e.g. waste, sludge, manure, straw, wood) and can be used for electricity generation through several processes (e.g. digestion, gasification, burning). The description starts with anaerobic biomass digestion, which was one of the first Dutch bioenergy technologies in the late 1970s; it subsequently experiences ups and downs. Next, we turn to waste incineration and co-firing of biomass and fossil fuels (coal and later also natural gas), which now provide the largest biomass contribution to renewable energy, but also face opposition from environmental and local interest groups.

Expectation dynamics of biomass digestion

Biomass digestion in an oxygen free environment produces biogas, which contains methane that can be used to generate electricity or heat. Visions on manure digestion were highly influenced by external factors and emerging problems within the agricultural and waste regime. In response to high oil prices in the late 1970s, a small network of farmers and agricultural researchers began to investigate small *farm-scale* manure digestion as a source of energy. Technical problems and (after 1986) decreasing oil prices lowered the expectations of this option.

By the mid 1980s, visions shifted towards large centralised plants. The functional focus was less on the production of biogas and more on manure processing (transforming it into less environmentally damaging substances). Expectations were very high, because centralised plants seemed a solution to the major manure surplus problems in the agricultural regime, but negative outcomes of learning processes created major disappointments a few years later, marking the end of the biogas hype. The backlash was that manure digestion was perceived as not feasible and manure digestion disappeared from the policy agenda.

In the late 1990s, interest in manure digestion was revitalised because of climate change and energy security concerns. Product champions pointed to the successful application of biomass digestion in Denmark and Germany, where manure digestion was developed as a multi-functional technology targeting problems in agriculture, energy and waste regimes (Geels and Raven 2006). Despite the new enthusiasm, total contributions to renewable energy are still small.

Social network

Farm-scale manure digestion in the late 1970s was pioneered by actors outside the energy regime. Projects were set up by agricultural researchers and individual farmers. Research was supported in the framework of a national research program, funded by the government.

In the mid 1980s, when centralised plants were seen as solution to manure surplus problems, more powerful agricultural regime actors took over (e.g. national farmers' organisations, the Ministry of Agriculture, national banks). These regime actors were under much societal and political pressure because of manure problems. The manure digestion network crumbled in the mid 1990s, because of negative experiences and costly failures.

For some years, no activities took place. In the late 1990s, however, manure digestion re-emerged with research institutes, consultants, energy distribution companies and several technology suppliers forming a new advocacy coalition (Raven 2005).

Learning processes

Experimental projects in the late 1970s and early 1980s encountered many technical and economic problems (e.g. sulphur contaminations that damaged gas turbines, lower biogas yields than expected). Decreasing oil prices after 1986 further damaged the economic viability of farmscale manure digestion, leading to a decline in projects (Geels and Raven 2006).

In the mid 1980s, the focus shifted to centralised biogas plants and manure processing. A large plant was constructed in 1988 (Promest), which was about 15 times larger than previous farmscale plants. This rapid upscaling created a range of technical and operational problems. Nevertheless, and despite warnings from consultants, the plant was enlarged with a factor 6 in 1990. Investment costs were €100 million, 40% of which came from investment grants from the Ministry of Agriculture. The enlarged plant opened in 1993. But technical and commercial problems rapidly became so large that the plant was closed in late 1994 (Geels and Raven 2006).

Meanwhile, German and Danish actors experimented with co-digestion, i.e. the addition of biomass sources (e.g. fatty wastes) to manure. This resulted in higher biogas yields and in technically stable and economically feasible biogas plants (Geels and Raven 2006). In the late 1990s, Dutch product champions pointed to these positive learning processes, arguing that biogas plant were multi-functional with environmental benefits for the energy, agriculture and waste domains. Policy makers also developed new interests in manure digestion, but the addition of waste products created legislative and permit problems (due to conflicting regulations for waste and agriculture). Legislative changes in the early 2000s led to new manure digestion projects.

Expectation dynamics and social network of waste and co-firing

Energy from waste has long been an important alternative energy source, although the government only recognised it officially as a sustainable source in the 1990s. In the early 1980s the government made a distinction between waste (like sludge and demolition wood) and biomass (cultivated energy crops). As energy from waste grew rapidly during the 1980s this distinction became blurred in official reports. Biomass was increasingly considered to include almost any organic fuel resource.

Waste was attractive to energy regime actors, because it offered them a relatively inexpensive fuel; utilities could earn 'gate fees' that would otherwise be paid for the dumping of waste streams.

In the early 1990s, utilities introduced co-firing of biomass and fossil fuels on a small-scale (Raven 2006). An exemption from the energy tax, introduced in 1996, made biomass co-firing financially even more attractive for the energy companies (Hofman 2005). Another driver for co-firing was pressure from the government. Increasing concern over climate change translated into stricter policies from the minister of Environmental Affairs. He threatened to close down coal-fired plants or to force a switch to natural gas. To avert this threat and reduce CO_2 emissions, utilities increased the level of biomass in coal-fired plants. Because of these activities, co-firing rapidly increased after 2000, becoming the largest renewable energy option (Figure 3).

Protests against these developments came from environmental groups who contested the official vision on biomass in the 1990s. They argued that waste was not sustainable and should be distinguished from biomass, which could be a sustainable form of energy. Environmentalists argued that plants that co-fired waste should fall under the stricter regulations from the waste regime. The utilities argued that waste in co-firing plants should be considered as biomass, which implied that certain emissions were allowed to be higher than for pure waste incinerators. These conflicting views led to legal struggles around permit procedures for new co-firing plants (Raven 2005). These legal procedures caused delay, and courts often ruled in favour of the plaintiffs. Nevertheless, the government broadened the definition of biomass to include all organic streams that can be used for energy production, including manure, sludge and demolition wood. Even the organic part of waste combusted in incinerators was (financially and statistically) recognised as a renewable energy source (Raven 2007). In the Third Energy White Paper (1995), the government expressed the vision of biomass as the main renewable option for the short- and mid-term.

Learning experiences

The expansion of co-firing was possible as a result of technical learning, which was relatively easy because regime actors could build upon existing competencies with coal combustion. Positive experiences showed that more biomass could be added than was initially expected, in particular when making use of more advanced technologies such as gasification. Although expectations about this more advanced technology were high, learning processes in the mid 1990s at a concrete biomass gasification plant (Amer plant), were negative. Despite a major re-design, the technical problems were not solved sufficiently (Raven 2006). Hence, regime actors turned back to more incremental ways of co-firing. Still, problems and uncertainties remained, for instance about the availability of biomass, about the development of a biomass market, about the impact of co-firing on the life span of power plants, and about the best technological route (Raven 2005). Recently, environmental organisations expressed concerns about two other issues: (1) competition between use of biomass for food production and for energy production and, (2) negative environmental impacts of large scale cultivation of energy crops, particularly in developing countries. Policy changes in subsidy schemes also contributed to uncertainty and, as a result, a sharp decline of co-firing in 2006 (Figure 2). This decline indicates that the commitment (and lock-in) from utilities to co-firing is not complete. Because co-firing is a dual-use technology, utilities can easily abandon biomass and switch back to normal practice if conditions become unfavourable.

2.3. *Fuel cells and hydrogen*

Fuel cells and hydrogen have been promising options for quite some time, with expectations displaying a hype cycle pattern. Fuel cells received much attention after successful application in the international space programs (1960s). In the 1970s and early 1980s, hydrogen came to be seen as an alternative energy carrier, but not as sustainable energy. More recently, the concept of

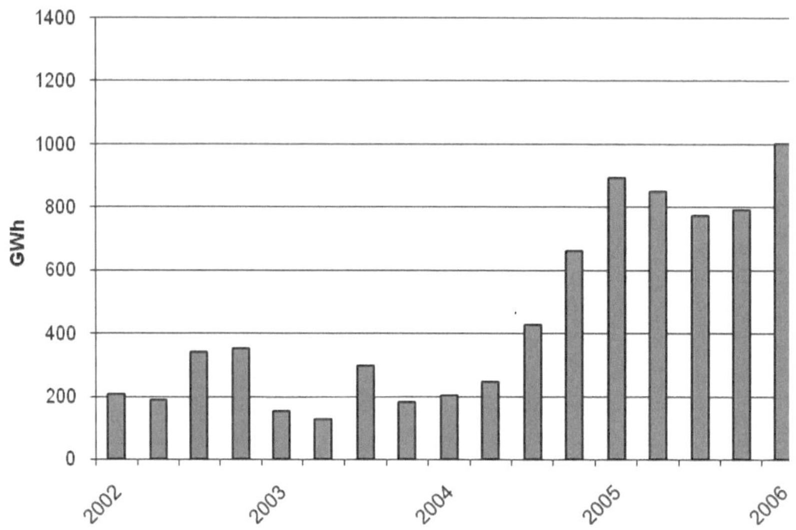

Figure 2. Development of co-firing of biomass 2002–2006 (http://statline.cbs.nl).

a hydrogen economy has become popular, because of its promise to solve many energy problems, in particular when the hydrogen is produced with renewable energy sources (Romm 2005).

Expectation dynamics

Interest for fuel cells emerged in the Netherlands in the early 1980s, when energy prices were high. International development efforts increased substantially in the USA and Japan. Pointing to a large US fuel cell demonstration project, industrial actors and researchers convinced the Ministry of Economic Affairs to start a dedicated R&D program to build up a new industry around fuel cells, which potentially had high conversion efficiencies and a lack of polluting waste products. In 1984, the Molten Carbonate Fuel Cell (MCFC) was selected as the most promising option. High expectations about this technology and the promise that a research program might help create a new Dutch industry were crucial elements in legitimating the funding for the second-largest energy R&D programme in the 1980s and 1990s (besides coal gasification). With regard to commercial prospects, the proponents and product champions promised that the MCFC project would deliver a competitive technology within 15 years. These promises were repeated regularly during the project, although outside consultants warned repeatedly about the lack of market demand, but the product champions went ahead anyway, pointing to international developments and the need to keep up. The Dutch government spent about €100 million, providing about 80% of total fuel cell funding. Despite these investments, the fuel cell innovation trajectory failed in the mid 1990s (Van der Hoeven 2001).

An important reason for this failure was that the vision about the main application domain changed from power generation to transport applications (fuel cell vehicles). In response to climate change pressures, car manufacturers became more interested in hydrogen powered cars, boosting new expectations. This change in vision halted MCFC development, because its technical characteristics (high operating temperatures) did not fit well with this new envisaged application domain. However, this new vision also created new opportunities for other fuel cell designs (e.g. the Proton Exchange Membrane FC (PEMFC), the Solid Oxide FC (SOFC)), which also benefited from broader enthusiasm about the hydrogen society.

Social network

Although there had been some previous R&D on fuel cells at universities and companies, a new network was build up in the 1980s. Although R&D activities were concentrated at ECN, the Ministry insisted on the participation of industrial partners. The Dutch steel company Hoogovens was enrolled in the network, but quit in 1988. Subsequently, two medium-sized companies were enlisted, because of their expertise in engineering and machine building (Schaeffer 1998). Also potential users were included in the network: the joint organisation of Dutch electricity producers (SEP) as 'launching customer' and EDCs as partners, because they saw fuel cell powered plants as an attractive option for small-scale decentralised generation (up to a few MW).

With regard to funding it seems that the Ministry of Economic Affairs was somewhat naive about the commitment and participation from industry. As soon as industry showed some interest, the Ministry provided subsidies (about 80% of total costs in this case), but real commitment was not very high, because of problems in the social network. The Ministry basically forced the two medium-sized companies to join after Hoogovens left. Under pressure, these companies participated half-heartedly. Furthermore, relations with potential users (especially energy distribution companies) also soured during the project. They were disappointed that the project did not deliver a product they could implement (Van der Hoeven 2001).

Learning processes

Despite some initial successes, learning processes advanced slower than expected. There were persistent technical problems: because of high operational temperatures of the MCFC (800–1000 °C) and the chemical substances used, cell components suffered corrosion. Because of these challenges, the program focused on technological learning, paying limited attention to market prospects. Technical learning and R&D activities (at ECN) focused primarily on the fuel cell itself, and neglected other components of a fuel cell system. This led to problems when researchers eventually assembled a system and learned that several design specifications of the fuel cell negatively impacted the whole system, making it very large and expensive. The project partners concluded that this concept would not produce a commercially viable product. In 1993 the project was cancelled, because there was no money or time to select another design (Van der Hoeven 2001). The EDCs were disappointed and tried to settle financial arrangements legally, but the SEP was satisfied, because they wanted to know if the MCFC was an option for large scale generation. The project showed that fuel cell power stations were technologically feasible, but economically not.

The project was restarted within 2 years, because ECN (and the Ministry) still believed in a future for the MCFC technology. The focus shifted to smaller systems for niche applications and, based on previous lessons, more attention was paid to commercial requirements (Van der Hoeven 2001), but in 1998, the project was again cancelled, signalling the end of MCFC research. Although the MCFC made good technical progress, it was increasingly seen as obsolete as expectations shifted to other fuel cell types, in particular the PEMFC and the SOFC. The main reason was a change in vision and economic opportunities (shift from power generation to transport applications) (Van der Hoeven 2001; Van den Hoed 2004).

2.4. *Photovoltaic (solar PV)*

Because of the small scale character of photovoltaic applications and because of the Dutch geographical position and climate, PV initially only received interest from researchers at universities.

In the early 1990s, perceptions of PV's potential improved substantially, but shifting policy priorities in the late 1990s caused stagnation in PV implementation, although societal support for PV is still high.

Expectation dynamics

In the 1970s, PV was not part of sustainable energy visions, because of the small scale of applications. Dutch policy makers and experts thought that PV was unfeasible in the Netherlands (Verbong 2001). In the early 1990s, however, official white papers on energy policy considered PV the most important renewable option for the long-term future. In the Third Energy White Paper (1995), for instance, the government denoted PV as the main long-term option, next to biomass. This remarkable shift was caused by active lobbying by researchers and positive learning processes (see below). Also, Shell, a powerful actor in the energy domain, embraced PV in its long-term scenarios (e.g. 'The Evolution of the World's Energy Systems', 1996), expecting it to be a major energy source by 2050. The Shell vision influenced policy makers and NGO (Verbong 2001).

Although visions and promises were high, electricity from PV remained more expensive than electricity generated by coal or gas fired plants and other renewable options. Technical development therefore focused on higher conversion efficiency of solar cells and lower production costs (Perlin 2002). Researchers promised that learning effects and economies of scale would eventually make this technology competitive with electricity from fossil fuel power plants. However, despite technological progress, the kWh price remained high. In the late 1990s, policy analysis re-evaluated available options for greenhouse gas reduction, using two criteria: (a) the contribution to greenhouse gas reduction, (b) cost-efficiency. PV scored poorly on both criteria: total contribution to renewable energy was small and it was very expensive. High PV expectations in the policy domain subsequently collapsed and the policy focus shifted back to wind and biomass.

Social networks and learning processes

Research on PV started at universities and research institutes. The high-tech character of the technology, closely related to the semiconductor technology, offered excellent opportunities for research (and funding). For a decade, the network remained relatively small, but sustained lobbying succeeded in persuading policy makers, industrial companies, utilities and environmental organisations to support research and demonstration projects (Verbong 2001). This facilitated learning processes, which demonstrated that PV could be applied in the Dutch climate. Projects abroad also contributed to more positive perceptions of PV in the early 1980s. By the late 1980s, there were efforts to develop grid-connected systems, but high costs provided important stumbling blocks.

In the early 1990s, the environmental movement embraced PV, because it had grown increasingly sceptical about other renewable energy sources. Broad societal support led to vastly increased budgets for R&D and PV pilot projects. Greenpeace, but also EDCs such as Nuon, started ambitious projects to encourage the diffusion of PV panels among Dutch households (Verbong 2001).

Tensions between the government and Shell then soured the social network. Shell, which owned a small production facility (Shell Solar), wanted to expand its PV activities in the late 1990s. The government initially stimulated Shell with the transfer of knowledge and technology from Dutch universities and ECN. Nevertheless, Shell decided to build its new PV production plant in Germany, which offered better protective and stimulating policies. This was a major disappointment for the Ministry of Economic Affairs. This failure in industrial policy was widely criticised for low reliability, lack of long-term commitment to solar energy (as the Germans did), too rapid introduction of market pressures, and insufficient support and protection, which is required for radical innovations (Verbong 2001).

3. Analysis and conclusions

Niche-innovation trajectories have long histories (30–40 years) and are non-linear processes with many ups and downs. The previous descriptions showed a high degree of trial and error and 'muddling through'. 'Muddling through' does not mean that no progress can be made. In fact, the share of renewable electricity has risen substantially, from less than 1% of total electricity consumption in 1990 to 4.3% in 2004 and 6.6% in 2006 (see Figure 3 for data on total production). Initially, renewable electricity came primarily from burning waste. Gradually, wind has become more important, supplemented from 1998 onwards by the rapid expansion of co-firing of biomass and fossil fuels. The use of biomass accounted for 4.3% of the national electricity production in 2005, but there has been a sharp decline in production in the second half of 2006 (Figure 2).

Policy makers like to claim this growth as evidence of the success of their policies. The previous descriptions, however, demonstrated that such claims are too easy. Instead of effective policies, we found costly failures, hype cycles, changing visions and policy priorities, and limited learning capability. The following two sub-sections use the three niche-internal processes (expectation dynamics, learning processes, social network dynamics) to provide a more systematic analysis of recurring patterns and mechanisms in Dutch niche development trajectories and policy dynamics.

3.1. *Patterns in niche development trajectories*

A. *Expectations*

High expectations and backlash. All niche-innovation trajectories were characterised by high initial expectations, e.g. wind energy in the 1970s, fuel cells in the mid 1980s, and PV in the

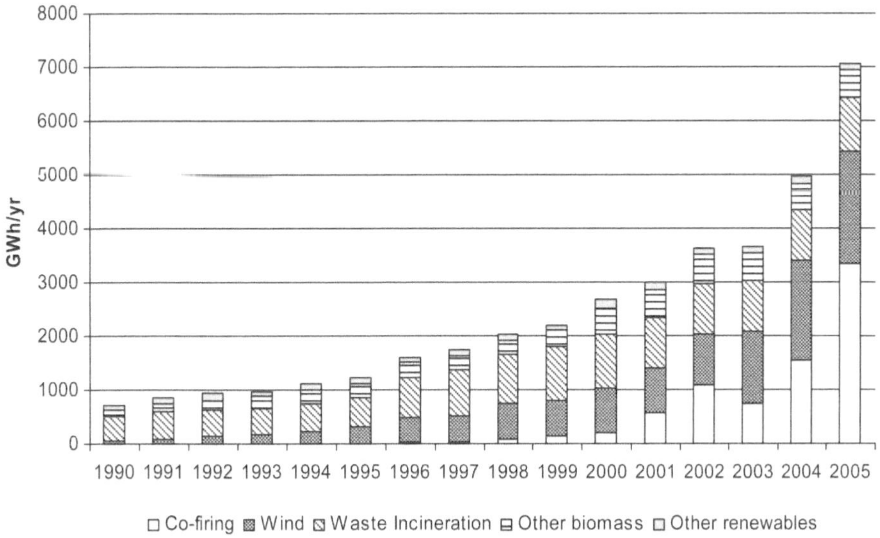

Figure 3. Dutch renewable electricity production (GWh per year) (data from Central Bureau of Statistics, www.cbs.nl).

early and mid 1990s. Product champions made high promises early in the trajectories to attract attention and funding. The technological visions triggered enthusiasm for some time, but were subsequently followed by disappointments. The general pattern is that expectations were too high, technical problems were bigger than expected, and implementation was slower than anticipated. So, most niche trajectories experienced a hype–disappointment cycle. This is not uncommon for innovation journeys, but our impression is that the peaks and valleys are more extreme than in other countries (see, for instance, the more gradual innovation journeys in Denmark (Garud and Karnoe 2003).

Expectation successions. Niche trajectories succeeded each other in the sense that disappointing outcomes and 'negative' learning processes in one trajectory often led to increased expectations in another trajectory. In the early 1990s, there was a shift from wind energy to biomass (for the short term) and PV (for the long term). In the early 2000s, there was a shift back from PV to wind energy.

This expectation dynamic also works in the other direction. Positive outcomes and learning processes in one trajectory may negatively influence the expectations of another trajectory. The present success of direct co-firing of biomass may become a barrier for the future development of more advanced technologies as gasification or pyrolysis. Figure 4 presents a schematic representation of expectation successions.

The mechanism behind this pattern is that policy actors face credibility pressures with regard to societal problems, such as sustainability. In societal debates they need to express faith in at least one promising solution. Policy makers are modernist actors who need to 'tell themselves forward'. They need to tell stories and make promises about how they will solve problems. This credibility pressure creates willingness to accept certain promises from product champions. It also explains why warnings about feasibility are often downplayed.

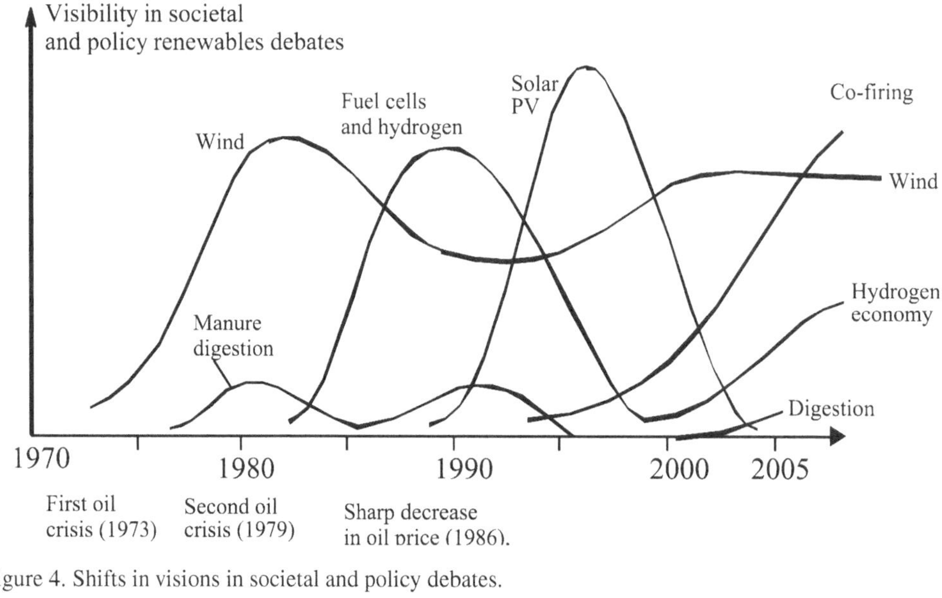

Figure 4. Shifts in visions in societal and policy debates.

B. *Learning processes*
Low quality learning processes. In most trajectories technological problems were underestimated. Regarding the molten carbonate fuel cell (MCFC), for instance, the problem of corrosion was much more persistent than the researchers wanted to believe, and regarding wind turbines, the transfer of knowledge from the aeronautic domain did not sufficiently take into account the much higher strain on rotor blades, leading to a string of broken blades. Although such problems are quite common with new technologies, they were insufficiently anticipated.

Technology push and focus on technological learning. All niche-innovation trajectories were characterised by a technology push approach, focusing primarily on R&D. Other dimensions of learning and articulation processes were relatively neglected (e.g. commercial prospects, societal embedding, legal procedures, societal stakeholders). This neglect negatively influenced subsequent implementation. Regarding MCFC, for instance, warnings about the lack of commercial feasibility were consistently ignored, contributing to the failure of the first phase. Lack of attention for the social acceptance of renewable energies led to major problems with wind turbines and biomass, which both experienced fierce opposition.

From 'stretch' to 'fit'. The growth in renewable electricity comes from options that are relatively close to the existing regime (have a 'fit' with existing rules, principles and practices). Innovation in the 1980s had more of a 'stretch' character, pioneering options that deviated substantially from the existing regime (and were more radical in that sense), but the late 1990s and early 2000s saw a move in the direction of 'fit' options that are closer to the existing regime. This applies not only to co-firing of biomass, but also to the CO_2 capture and sequestration that currently receives much attention. Regarding biomass, there were more radical technologies (like gasification), but these typically experienced more problems.

C. *Social networks*
Supply-side oriented innovation networks. Dutch energy innovation journeys have been dominated by a limited number of knowledge institutes (ECN, universities), large companies (Shell, utilities) and intermediaries (Senter-Novem). This composition of social networks, and the focus on R&D, led to the technology push approach, discussed above. This social structure seems to be a persistent feature of Dutch energy innovation. Although policy makers have adopted a discourse on innovation systems (which includes phrases such as 'user–producer interactions' and 'multi-directional knowledge flows'), real innovation practices still seem to be mainly R&D oriented. The government–industry–university relations have characteristics of 'old boy networks', which divide subsidies among themselves. For small firms, who are not part of the relevant networks, it is usually more difficult to get funding.

Narrow, closed networks. For most of the period, the social networks behind niche-innovations were narrow and closed. Outsiders were hardly involved. This was partly related to characteristics of Dutch innovation journeys discussed above, but technical characteristics also played a role. Photovoltaics, for instance, had a 'high tech' character, giving researchers a strong position in the PV network. Also modern wind turbines are 'high tech', building on knowledge from the aeronautical domain.

The main exception to this pattern was wind energy in the early 1980s, where a grass roots movement succeeded in setting up a decentralised research line in the national research programme. Local groups of enthusiasts also founded wind energy co-operations. Another exception is manure digestion, which was pioneered by farmers, researchers and technology suppliers in the agricultural domain, but renewable energy contributions from digestion have remained very small.

Network tensions. Although the social networks were narrow and closed, they were not harmonious, but strained. Technology developers and utilities clashed repeatedly. Government actions, which sometimes forced companies to participate in networks, also created tensions and damaged the willingness to cooperate (see also below).

Societal protest and contestation. Although 'renewable energy' as an abstract term is widely supported, concrete options are usually contested. Each renewable option has some disadvantages ('horizon pollution', 'bird shredding', emissions of heavy metals, stench, costs). Because Dutch institutional arrangements give stakeholders many options to protest, environmentalists and local inhabitants can easily frustrate or delay options they dislike. Wind energy and biomass projects experienced severe delays because of protests and legal procedures. Contestation was exacerbated by the technology push approach, which gave little attention to the societal embedding of new technologies. Because of the disappointing results during the 1990's, there appears to be more attention for stakeholder participation, which involves opposing groups earlier in the innovation process.

3.2. *Conclusions about policy*

With regard to Dutch energy policy, several patterns can be observed.

Link with industrial and innovation policy. Energy policy has been strongly linked to industrial and innovation policy. In all niche development trajectories industrial arguments were prominent. Wind energy, PV, and fuel cells were (at least partly) stimulated to support existing industry (Shell) or build new industries. These industrial interests have been important arguments for high levels of public funding of energy research. The exceptions were some biomass applications, which were either taken up by utilities themselves (and supported with other policy measures) or not perceived as part of the energy domain (biogas, which was sponsored by agricultural actors). The link between energy policy and industrial policy also occurred in other countries, where it had more success because of different innovation styles, e.g. more gradual upscaling in Denmark (Garud and Karnoe 2003; Kamp, Smits, and Andriesse 2004) and more stable protection and market creation in Germany (Jacobsson and Lauber 2006).

Instability and unreliability of policy. An important factor in Dutch industrial failures was policy instability. With regard to renewable energy, frequent changes occurred in regulations and subsidies, exemplified by the REB and MEP-regulations from 1995 onwards. Shifts in policy, almost every few years, created uncertainties and hampered private investments.

Low quality of network management. One problem was that firms were sometimes 'forced' to cooperate in projects, e.g. in the fuel cell project, leading to low commitments. Another problem is that policy makers were eager to accept promises from industry about commitment, technical feasibility, risks and future price/performance (in the case of fuel cells and PV), but policy makers appeared to have little competence to assess the robustness and quality of visions and promises. Money was made available, as soon as industry showed some interest. This seems to be a common characteristic of all large energy technology projects in the Netherlands (Verbong 2001). Industry does not invest in high risk, long-term technological projects, unless the government carries the financial risks, but thorough quality checks on technological promises were often absent.

Limited learning. Experiences in the 1990s and in other countries (Denmark, Germany) showed that regulatory stability and long term support were crucial for successful development of renewable technologies (Jacobsson and Lauber 2006). Policy analysts and consultants recognised these characteristics as key components for renewable energy policy, but these lessons were not internalised, as policy actions in the early 2000s demonstrate (frequent policy changes that created instability and uncertainty). Policy makers did not provide long-term guarantees and too rapidly abandoned innovations when setbacks occurred. They cut off protection measures too soon, and introduced market pressures when innovations had not yet stabilised. So, the lessons from the 1990s were either forgotten or not implemented (for political or financial reasons).

Bandwagon and circus dynamics. Another reason for policy instability is that renewable energy is highly politicised. Policy makers in this domain are eager to show successes and are therefore looking for positive stories. This enhances the pattern of expectation successions, described above. Policy makers jump on the bandwagon and thus reinforce hype cycles, but when setbacks occur a few years later, they become disappointed and the circus moves on to a new promising option. This gives Dutch innovation a hype-driven character, and makes policy unreliable (a common complaint by innovators). Changes in subsidies or regulations directly affect the feasibility of experiments and implementation projects.

These recurring characteristics of innovation journeys and policy combine into a Dutch renewable energy 'innovation style'. Compared to other countries such as Denmark and Germany, this style has not been very successful in nurturing the development of radical options since the 1970's. The renewable electricity expansion in the first years of the 21st century was carried mainly by options that are closer to the existing regime (and hence more incremental). We conclude that the Dutch innovation style may be substantially improved with regard to learning processes (paying more attention to social aspects), the construction of broader networks (including users and outsiders like NGOs and SMEs) and expectations (more critically assessing the promises articulated by the representatives of the vested interest and technology promoters).

Note

1. See the special topic forum on 'Theory development: Evaluation, reflections, and new directions' in the *Academy of Management Review* 24, no. 4 (1999); special issue on 'Processual research on management and organizations' in the *Scandinavian Journal of* Management 13, no. 4 (1997); special issue on 'Longitudinal field research methods for studying processes of organizational change' in *Organization Science* 1, no. 3 (1990).

References

Abbott, A. 1992. From causes to events: notes on narrative positivism. *Sociological Methods and Research* 20, no. 4: 428–455.

Bergek, A., and S. Jacobsson. 2003. The emergence of a growth industry: a comparative analysis of the German, Dutch and Swedish wind turbine industries. In *Change, transformation and development*, eds. S. Metcalfe and U. Cantner, 197–228. Heidelberg: Physica/Springer.
Bijker, W.E., T.P. Hughes, and T. Pinch, eds. 1987. *The social construction of technological systems: new directions in the sociology and history of technology*. Cambridge, MA: The MIT Press.
Breukers, S. 2007. Changing institutional landscapes for implementing wind power. PhD thesis, University of Amsterdam.
Brown, N., and M. Michael. 2003. The sociology of expectations: retrospecting prospects and prospecting retrospects. *Technology Analysis & Strategic Management* 15, no. 1: 3–18.
Garud, R., and P. Karnoe. 2003. Bricolage versus breakthrough: distributed and embedded agency in technology entrepreneurship. *Research Policy* 32: 277–300.
Geels, F.W., and R.P.J.M. Raven. 2006. Non-linearity and expectations in niche-development trajectories: ups and downs in Dutch biogas development (1973–2003). *Technology Analysis & Strategic Management* 18: 375–92.
Griffin, L.J. 1993. Narrative, event-structure, and causal interpretation in historical sociology. *American Journal of Sociology* 98, no. 5: 1094–133.
Hofman, P.S. 2005. Innovation and insitutional change. The transition to a sustainable electricity system. PhD thesis, University of Twente, Enschede.
Hoogma, R., R. Kemp, J. Schot, and B. Truffer. 2002. *Experimenting for sustainable transport: the approach of strategic niche management*. London: Spon Press.
Jacobsson, S., and V. Lauber. 2006. The politics and policy of energy system transformation – explaining the German diffusion of renewable energy technology. *Energy Policy* 34, no. 3: 256–76.
Kamp, L.M., R.E.H.M. Smits, and C.D. Andriesse. 2004. Notions on learning applied to wind turbine developments in the Netherlands and Denmark. *Energy Policy* 32: 1625–37.
Kemp, R., J. Schot, and R. Hoogma. 1998. Regime shifts to sustainability through processes of niche formation: the approach of strategic niche management. *Technology Analysis & Strategic Management* 10: 175–96.
Mayr, E. 1963. *Animal species and evolution*. Cambridge MA: Harvard University Press.
Ministry of Economic Affairs. 1995. *Third energy white paper*. The Hague: SDU.
Ministry of Spatial Planning, Housing and the Environment. 1989. *The national environmental policy plan*. The Hague: SDU.
Pentland, B.T. 1999. Building process theory with narrative: from description to explanation. *Academy of Management Review* 24, no. 4: 711–24.
Perlin, J. 2002. *From space to earth. The story of solar electricity*. Cambridge, MA: Harvard University Press.
Pettigrew, A.M. 1990. Longitudinal field research on change: theory and practice. *Organization Science* 1, no. 3: 267–92.
———. 1997. What is a processual analysis. *Scandinavian Journal of Management* 13, no. 4: 337–48.
Raven, R.P.J.M. 2005. Strategic niche management for biomass. PhD thesis, TU Eindhoven.
———. 2006. Towards alternative trajectories. Reconfigurations in the Dutch electricity regime. *Research Policy* 35: 581–595.
———. 2007. Co-evolution of waste and electricity regimes. Multi-regime dynamics in the Netherlands (1969–2003). *Energy Policy* 35: 2197–208.
Romm, J.J. 2005. *The hype about hydrogen. Fact and fiction in the race to save the climate*. Washington, DC: Island Press.
Schaeffer, G.J. 1998. Fuel cells for the future: a contribution to technology forecasting from a technology dynamics perspective. PhD thesis, Twente University, Enschede.
Shell. 1995. *The evolution of the world's energy system 1860–2060*. London: Shell International.
Schot, J.W., and F.W. Geels. 2007. Niches in evolutionary theories of technical change: a critical survey of the literature. *Journal of Evolutionary Economics* 17, no. 5: 605–22.
Van den Hoed, R. 2004: Driving fuel cell vehicles. How established industries react to radical technologies. Thesis, Delft.
Van der Hoeven, D. 2001. *Een Gedurfd Bod: Nederland zet in op de Brandstofcel*. Bergen NH: BetaText.
Van de Ven, A.H., D.E. Polley, G. Garud, and S. Venkataraman. 1999. *The innovation journey*. Oxford: Oxford University Press.
Van de Ven, A.H., and M.S. Poole. 2005. Alternative approaches for studying organizational change. *Organization Studies* 26, no. 9: 1377–404.
Van Est, R. 1999. Winds of change. A comparative study of the politics of wind energy innovation in California and Denmark. PhD thesis, University of Amsterdam, Amsterdam.
Van Lente, H. 1993. Promising technology: the dynamics of expectations in technological development. PhD thesis, Twente University, Delft: Eburon.
Van Rooijen, S.N.M., and M.T. van Wees. 2006. Green electricity policies in the Netherlands: an analysis of policy decisions. *Energy Policy* 34: 60–71.

Verbong, G.P.J. 1999. Wind power in the Netherlands 1970–1995. *Centaurus* 41, no. 1/2: 137–60.

Verbong, G., A. van Selm, R. Knoppers, and R. Raven. 2001. Een kwestie van lange adem. De geschiedenis van duurzame energie in Nederland 1970–2000. Boxtel: Aeneas Technical Publishers.

Verbong, G., and F. Geels. 2007. The ongoing energy transition: lessons from a socio-technical, multi-level analysis of the Dutch electricity system (1960–2004). *Energy Policy* 35: 1025–037.

‘Legitimation’ and ‘development of positive externalities’: two key processes in the formation phase of technological innovation systems

Anna Bergek, Staffan Jacobsson and Björn A. Sandén

Responding to the climate change challenge requires a massive development and diffusion of carbon neutral technologies and, thus, emergence and growth of new socio-technical systems. This paper contributes to an improved understanding of the formative phase of new technological innovation systems (TIS) by outlining a framework for analysing TIS dynamics in terms of structural growth and key innovation-related processes (“functions”) and by discussing two of these functions at some depth: “legitimation” and “development of positive externalities”. Empirical examples are provided from case studies on renewable energy technologies. We highlight the problematic role of technology assessment studies in shaping legitimacy and the importance of early market formation for the emergence of “packs of entrepreneurs” that may contribute to legitimation, and discuss how exploitation of overlaps between different TISs may create positive externalities, opening up for a powerful “bottom-up” process of system growth. Associated policy and management challenges are identified.

1. Introduction

The literature on climate change makes it plain that much of the capital stock in the energy and transport sector needs to be replaced within three to four decades (Stern 2006). This requires a massive development and diffusion of carbon neutral technologies. To be able to facilitate such a transformation, policy makers and others need to understand the key processes in the development, diffusion and utilisation of new technologies. This appreciation is today partly lacking. This is particularly so for those informed by neoclassical economics, which is concerned with the selection of technologies ‘from the shelf’ rather than ‘filling the shelf’ with new technologies (Sandén and

Azar 2005). The latter process is synonymous with the emergence and growth of new socio-technical systems – a messy and difficult process that is yet not fully understood. The purpose of this paper is, therefore, to contribute to an improved understanding of the early and formative phase of new technological innovation systems (TIS) – and to identify associated policy and management challenges.

We begin, in Section 2, to develop an understanding of the dynamics of TIS, both in terms of structural growth and key innovation-related processes. We label these latter processes 'functions' of the innovation system and all of these are potentially linked to challenges for decision makers desiring to influence the formation of new TIS. In Section 3, we discuss at some length two of the all-in-all eight functions. These are 'legitimation' (acquiring a social acceptance of new technologies) and 'development of positive externalities' (that underpin complementarities of supposedly competing technologies). The focus on legitimation and external economies is chosen because of an apparent discrepancy between our empirical findings and the lack of attention given to these functions in the literature on technology management and policy. Empirical evidence is taken from case studies of TISs in Sweden and Germany, centred on three renewable electricity production technologies (biopower, wind power and solar photovoltaic power) and a number of alternative transportation fuels (e.g. methanol, ethanol, biogas, hydrogen) in Sweden. These case studies have been thoroughly described elsewhere and will not be detailed here.[1] Our main conclusions are given in Section 4.

2. The dynamics of technological innovation systems: an outline of a framework

A general definition of a system is a group of components (devices, objects or agents) serving a common purpose, i.e. working towards a common objective or overall function.[2] In studies of technical change, the concept of 'technological system' has been used in two different, but partly overlapping, ways. The first is a socio-technical system specialised on the provision of a particular output that serves some (societal) function, e.g. electricity production or transport (Hughes 1983, 1987; Wenk and J. Kuehn 1977). The second focuses on the development, diffusion and utilisation of new technologies (Carlsson and Stankiewicz 1991; Jacobsson and Johnson 2000), and may, thus, alternatively be labelled 'technological innovation system' (TIS). Arguably, these two concepts describe different aspects of the same phenomenon: Whereas the first captures all aspects of a 'technology delivery system' (Wenk and Kuehn 1977), the latter focuses on the innovative activities within it. In this paper, we use this more narrow system concept since we are primarily interested in the development and diffusion of new technical solutions.

The components of a technological innovation system are the actors, networks and institutions contributing to the development, diffusion and application of a particular technology (Bergek 2002; Carlsson and Stankiewicz 1991; Galli and Teubal 1997). In line with a number of previous authors (Geels 2004; Hughes 1983; Sandén and Jonasson 2005), we also include the technology as such among the components. Each TIS has a degree of uniqueness in terms of its structural components (i.e. some components are exclusively dedicated to the TIS), but components may also be part of several systems simultaneously (Wenk and Kuehn 1977). In emerging TIS, few specialised components can be expected to exist and the structural overlap with – and dependence on – other systems is, therefore, likely to be particularly large.

Technology includes both artefacts and knowledge (Das and Van de Ven 2000; Layton 1974). Artefacts may come in the form of hardware (e.g. products, design tools and machinery) or software (e.g. procedures/processes and digital protocols). Knowledge is partly found as competence within actors, but also codified in recipes (text, drawings, etc.) or embedded in artefacts. It may be of a more formal (research-based) character or of a more experienced based character (Asheim and Isaksen 1997).

Actors include firms (and other organisations) along the whole supply chain. For instance, in the case of solar cells they include manufacturing of machinery to make solar cells, cell and wafer producers, engineering firms designing and delivering whole systems, roof and façade manufacturers, electricians, architects and users. They also include universities, industry associations, bridging organisations, interest organisations such as Greenpeace, and government bodies. Although the mere existence of actors in a TIS may in some cases be of great importance for, e.g. legitimacy (Carroll 1997), the core of this component is agency.

The *institutional* base of the TIS regulates interactions between actors. Institutions may come in the form of hard regulations (controlled by juridical systems) and in the form of norms and cognitive rules (controlled by social systems). The normative and cognitive dimensions of a TIS are not always easily separated – actors are guided in what they think is right and what they want to do by what they know and are able to do. This interpretation is closely related to the concepts of 'limited (or bounded) rationality' (Simon 1955), 'bounded vision' (Fransman 1990) and 'path dependency' (David 1988). Taken jointly, normative and cognitive institutions influence decisions and actions in the form of frames (Bijker 1995; Geels and Raven 2006) or paradigms (Dosi 1982) that structure learning processes (problem agendas, ways to do business, etc.). An integral part of these frames are expectations, i.e. beliefs about the future (Van Lente 1993).

The formation of *networks* follows not automatically from the entry of various organisations into the TIS, but requires links that turn fragmented components into a system. Of particular importance here are different types of networks between actors. Learning networks link suppliers to users, related firms or competitors, or university to industry. They constitute important modes for the transfer, or sharing, of knowledge and also influence the perception of what is possible and desirable, which guides specific investment decisions (Carlsson and Jacobsson 1993; Geels and Raven 2006). Political networks are of equal importance. Policy making takes place in a context where advocacy coalitions, made up of a range of actors sharing a set of norms or beliefs, compete in influencing policy in line with those beliefs (Sabatier 1998). In addition to these actor networks, artefacts may be linked together in different ways. They may be hard-linked, to make up larger artefacts/product systems (e.g. components in a car), or more loosely linked, in infrastructural networks (e.g. cars and petrol stations).

When a radically new technology emerges, few of these structural components are in place. The TIS must, thus, go through a formative phase, in which the constituent elements evolve and agglomerate through entry of firms and other organisations, formation of networks, institutional alignment and the accumulation of knowledge and physical artefacts.[3] This is a cumulative process of many small changes (Van de Ven and Garud 1989), which can last for decades.[4] It is also a process characterised by high uncertainty in terms of technologies, markets and regulations (Kemp, Schot, and Hoogma 1998; Van de Ven 1993), which both investors and policy makers must handle.

Once the elements are 'put in place', the TIS is in a position to shift to a growth phase (Carlsson and Jacobsson 1997; Porter 1998). As and when it does so, a chain reaction of powerful positive feedbacks may materialise, setting in motion a process of cumulative causation (Myrdal 1957).[5] Indeed, virtuous circles are central to a development process – as they are formed, the evolution of the TIS becomes increasingly self-sustained.[6] Through further positive feedback processes, the TIS may eventually become mature, characterised by a stable structure (or even structural

Table 1. Functions of technological innovation systems.

Development of formal knowledge	The breadth and depth of the formal, research-based knowledge base and how that knowledge is developed, diffused and combined in the system.
Entrepreneurial experimentation	Knowledge development of a more tacit, explorative, applied and varied nature – conducting technical experiments, delving into uncertain applications and markets and discovering/creating opportunities, etc.
Materialisation	The development of (and investment in) artefacts such as products, production plants and physical infrastructure.
Influence on the direction of search	The extent to which supply-side actors are induced to enter the TIS, or put more subtly, direct their search and investments towards the TIS.*
Market formation	Articulation of demand and more 'hard' market development in terms of demonstration projects, 'nursing markets' (or niche markets), bridging markets and, eventually, mass markets (large-scale diffusion).
Resource mobilisation	The extent to which the TIS is able to mobilize human capital, financial capital and complementary assets from other sources than suppliers and users and the character of this mobilisation.
Legitimation	The socio-political process of legitimacy formation through actions by various organisations and individuals. Central features are the formation of expectations and visions as well as regulative alignment, including issues such as market regulations, tax policies or the direction of science and technology policy.
Development of positive externalities ('free utilities')	It reflects the strength of the collective dimension of the innovation and diffusion process. It also indicates the dynamics of the system since externalities magnify the strength of the other functions.

Source: Based largely on Bergek et al. (2008b).
*This function also covers the mechanisms influencing the direction of search *within* the TIS, in terms of different competing technologies, applications, markets, business models, etc.

'lock-in').[7] Such a system develops primarily through the forces of its own momentum and is quite resilient to external pushes and pulls.

To analyse this kind of dynamics we suggest that it is fruitful to distinguish a number sub-processes that are directly related to the innovation process, i.e. the development, diffusion and use of new products, processes, etc., and that influence the build up of system structures. In Table 1, we outline eight sub-processes and label these 'functions' (Bergek and Jacobsson 2003; Johnson 1998; Johnson 2001; Johnson and Jacobsson 2001). While the exact number of functions is somewhat arbitrary, we suggest that the set covers the key processes in the dynamics of any TIS. These functions have been derived from a rich and multidisciplinary theoretical base, including literature on innovation systems and evolutionary economics but also political science, sociology and population ecology (see Bergek et al. 2008a). Earlier versions of this framework have been applied in a number of empirical studies (e.g. Jacobsson and Bergek 2004; Negro, Hekkert, and Smits 2007).

In our previous research, we have used these functions to analyse the performance of different TIS (see, e.g. Bergek and Jacobsson 2003; Johnson and Jacobsson 2001). We have also shown how policy makers may assess the 'goodness' of a specific functional pattern and use it to guide

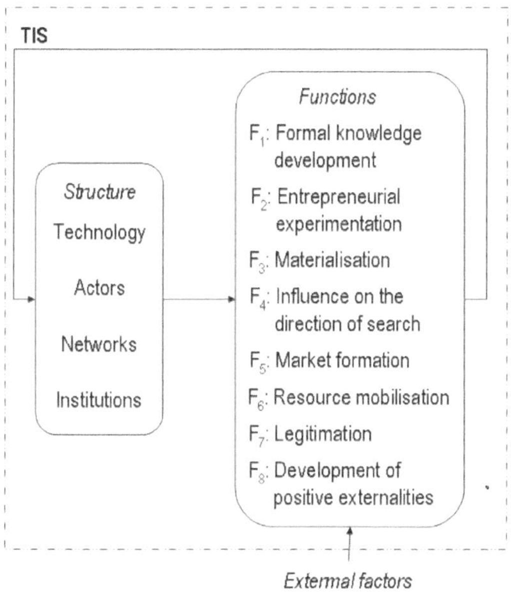

Figure 1. Relations between external influence, structural elements and functions.

their policies (Bergek et al. 2008a). In this paper, however, we focus primarily on the issue of TIS dynamics.

The strength of the functions is determined not only by the influence from structural components (endogenous dynamics) but also by exogenous factors at many levels, e.g. regime/landscape (Geels 2002) or national/sectoral innovation systems.[8] This co-existence of internal and external influences was highlighted already by Myrdal, who even stated that '... the main scientific task is ... to analyse the causal inter-relations within the system itself as it moves under the influence of outside pushes and pulls and the momentum of its own internal processes' (Myrdal 1957).

In an early phase of system formation, exogenous factors may dominate (Raven 2005), simply because the system components are underdeveloped. As the system grows, however, the endogenous dynamics may gain strength, involving extensive linkages between functions. Alterations in the functions impact on the structure of the system and an altered structure of the TIS will, in turn, influence the character of the functions (see Figure 1).

As an illustrative example of coupled functional and structural development, we will describe the emerging solar cell TIS in Sweden (see Palmblad et al. 2006). 'Legitimation' was shaped by (1) lobbying from political networks in an embryonic solar cell industry ([1] in Figure 2), (2) the climate change debate [2] and (3) experience from 'sister' TIS (e.g. that in Germany) [2]. The resulting changes in the institutional framework in terms of norms and regulation [3] in turn strengthened the functions 'influence on the direction of search' and 'market formation' [4], inducing a number of new firms and users to enter [5]. This strengthened the functions 'entrepreneurial experimentation' and 'materialisation' [6], also made possible by the availability of imported components [7], and led to structural change in terms of technology [8]. The existence of working and visible PV systems is likely to strengthen 'legitimation' [9]. Some of the firms recognised the need, and argued for, collective action and formed an industry organisation for that purpose. In the future, this may improve the process of legitimation further [10] (complete

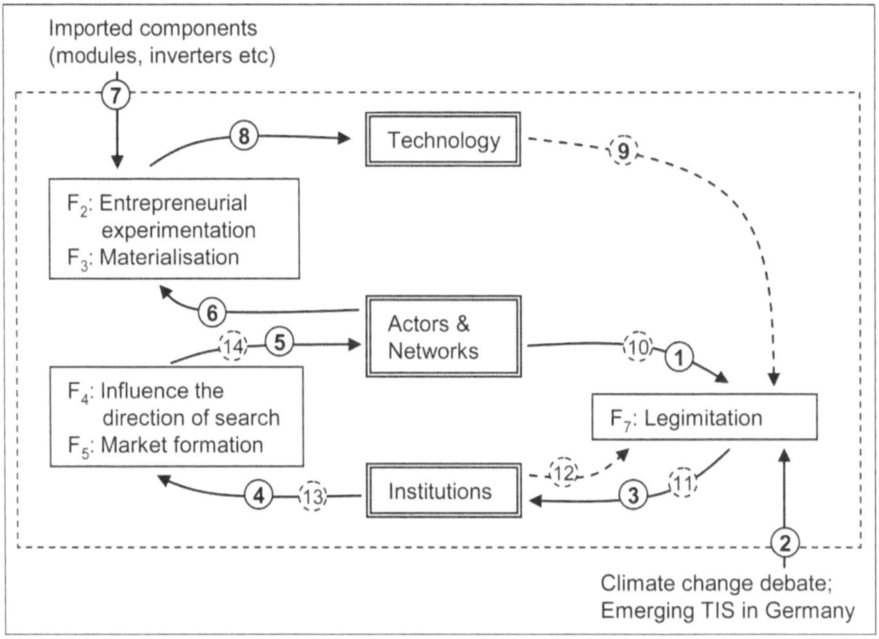

Figure 2. Example of functional and structural development (the solar cell TIS in Sweden).

feedback loop), inducing a favourable regulatory framework [11]. This institutional alignment may then stimulate further 'legitimation' [12] and 'influence on the direction of search' [13] of yet more entrants [14].

From this framework follows that decision makers (management, policy, interest organisations) with an ambition to influence system dynamics may acquire an understanding of the micro level processes involved by capturing the strength and nature of each of the functions, how they interact with each other and influence the structural components, enlarging the TIS. They may then pinpoint any of these functions for system-building activities. To the extent that such intervention is successful, the structural elements of the TIS will alter, which then opens up for new opportunities for agency to intervene.

3. Legitimation and development of external economies in the formative phase

In this section, we will explore more deeply two of the functions in Table 1, 'legitimation' and 'development of positive externalities'. This does not imply that we consider the other functions to be less important. Nor does it imply that other functions are not present in the descriptions, since, as pointed out above, the functions are closely linked and coupled in cause-effect chains. Rather, we 'look into' the system through the two selected functions. The focus on legitimation and external economies is chosen because of a perceived discrepancy between our empirical findings and the received literature.

First, whilst a number of studies on new technologies in general – and renewable energy technologies (see also the contribution by Agterbosch and Breukers in this special issue) and transport fuels in particular – underline the centrality of the process of legitimation (Aldrich and Fiol 1994;

Deeds, Mang, and Frandsen 2004; Rao 2004; Van de Ven 1993), it is absent in much of the literature on 'Policy and Innovation Systems' (Bergek et al. 2008b). Perhaps acquiring legitimacy is particularly problematic in the energy and transport sectors, where incumbent technologies, actors and institutions have had decades (in some cases more than a century) to secure their positions as familiar and socially accepted. Incumbents are, indeed, powerful and highly organised and protect their investments – for which they receive very large direct and indirect subsidies – against discontinuous changes. Incumbents are, thus, not only hesitant towards adopting the new technologies (Hellsmark 2004; Teppo 2006), but may also deliberately attempt to block the development of new TIS (Bergek 2002; Kaiser 1992). Much of the early efforts of an emerging TIS thus have to be spent on legitimation issues.

Second, as regards the 'development of positive externalities', we will take a particular angle by focussing on positive externalities that flow between *different* TIS. Conventional management and policy literature assumes a competitive relationship between technologies whereas our previous empirical work reveals a more complex picture including also strong complementarities. The focus on competition is prominent both in the management literature, with its focus on 'competing designs' in early phases (Utterback 1994), and in the policy literature, which emphasise the value of regulatory frameworks leading to competition between technologies so as to secure that the most cost efficient solution is selected (e.g. Alfsen and Eskeland 2007). Shifting attention from competition only, to include finding common grounds on the basis of which advocates of various technologies may seek to collaborate could be particularly interesting in the context of climate change, which necessitates the parallel development of many TIS centred on different carbon neutral technologies (Stern 2006).[9]

3.1. *'Legitimation'*

Legitimacy is a matter of social acceptance and compliance with relevant institutions. It is necessary to attain legitimacy in order for resources to be mobilised, for demand to form and for actors in the new TIS to acquire political strength (e.g. Aldrich and Fiol 1994). Thus, and as is widely acknowledged in previous research, legitimacy is a prerequisite for the formation of new industries and, we would add, new TIS (e.g. Aldrich and Fiol 1994; Rao 2004; Van de Ven 1993).

Legitimacy is, however, not given but rather formed through conscious actions by various organisations and individuals in a socio-political process of legitimation, which incorporates cognitive, normative as well as regulative aspects. The most commonly described strategy for industry legitimation is to conform to established institutions (Oliver 1991; Suchman 1995). For example, suppliers of ethanol fuel for cars may decide to advocate an increased use of low-ethanol blends for gasoline cars, which would imply regulative legitimation in terms of following the European Commission's fuel specifications for gasoline and normative legitimation in terms of allowing customers to remain true to their refuelling habits as well as their car preferences. However, in many cases of low-carbon innovation, existing institutions tend to block the development of new technological options (Unruh 2000). The strategic option (for firms) is, then, the most difficult one: to influence the institutional framework in order to achieve institutional alignment (Oliver 1991; Zucker 1987). This may involve manipulation of established institutions (in order to bring them into conformity with the emerging TIS) or the establishment of new institutions (cf. Ashforth and Gibbs 1990; Oliver 1991; Suchman 1995; Zimmerman and Zeitz 2002).

In the earliest stage of the formative phase, legitimation mainly involves getting the technology accepted as a desirable and realistic alternative to incumbent substitutes.[10] In essence, legitimation then becomes 'the politics of shaping expectations and of defining desirability'.[11] Before the components of the emerging TIS have been put in place, this is primarily done through 'expert' legitimation built on technology assessments (TA) and 'rational' arguments. As the proponents of the new TIS gain strength and political power, legitimation built on the familiarity, commitment and trust created by the accumulation of actors will increase in importance. In the following, we will discuss these two types of legitimation in more detail (for other aspects of legitimation, see the contribution by Agterbosch and Breukers in this special issue).

Early in the formative phase, when there is little experience of the technology, TA studies supply information about performance and potential. This is an 'expert' legitimation based on 'rational' arguments grounded in the knowledge base and problems articulated at a societal level. The role of assessment is, however, not unproblematic. It may cause genuine uncertainty for investors and policy makers and also create an opportunity for various actors (in different advocacy coalitions) to select assessment designs to legitimise or delegitimise a technology in order to serve their own 'political' goals. We will discuss three sources of these problems.

First, *changing problem agendas* at the societal, or landscape, level may alter the weight of different performance criteria and the power of different arguments. Sandén and Jonasson (2005), for instance, describe the shift of external driving forces for the development of alternative fuels in Sweden from fear of oil depletion (1974 and 1985), through local air quality (1985–mid 1990s), to climate change and a return to the fear of oil depletion. The shift in the 1980s had a dramatic effect on the legitimacy of methanol (from gasified feedstock), which was replaced by ethanol and methane as main alternatives. The second shift resulted in a move back towards gasified biomass (and towards hydrogen). What is legitimate, may, thus, shift rapidly in response to changing problem agendas.

Second, uncertainty may stem from *differences in the weight of various performance criteria* between actor groups, guided by diverging institutions. Because discontinuous technological change often implies substantial uncertainty with regards to future relevant systems-in-use and value networks (Rosenbloom and Christensen 1994), it may be difficult for entrepreneurial actors to choose which institutional framework to adapt to. For example, in the case of black liquor gasification in Sweden, an attempt by entrepreneurial actors to win legitimacy in the eyes of incumbent utilities and government policy makers by adapting the technology to the current electricity output maximisation norm in *that* field was in vain. Potential users in the paper and pulp industry instead favoured technology features improving the quality of the pulp and the flexibility of the pulping process (Bergek 2002).

Third, even if there are shared performance criteria, assessments may give varying results because of *methodological choices* that are subtle and difficult to unmask for non-specialists. For this reason, assessments are often used in early institutional battles as a way to promote a specific technical solution. In the energy sector, this 'politics of expectations and desirability' is, in particular, reflected in widely diverging statements of the potential of various renewable energy technologies.[12] In the Swedish wind power case, a potential of 20 TWh or more has been put forward by wind power proponents since the 1970s. However, a scrutiny of the major Swedish daily newspapers for articles that mention wind power reveals that more than 80% of these described the potential or actual wind power contribution as 'small' or 'smaller than some other electricity source'. These negative expectations on wind power, communicated through media, have not been counteracted by policy makers in a forceful way and this lack of a national vision for wind power has clearly contributed to a low legitimacy for wind power in the eyes of important stakeholders (Bergek 2002).

The wind power potential is mostly described in terms of a comparison with nuclear power. This is symptomatic for the decades long Swedish 'nuclear power trauma', during which both wind power and biopower have been met with resistance from advocates of nuclear power.[13] A particularly telling example is an editorial in the leading (emphatically pro-nuclear) daily newspaper *Dagens Nyheter*, which claimed that:

> Firewood can't replace nuclear power . . . Biofuels can give us a marginal addition, but not even with maximum production can they, with respect to power, replace more than one nuclear power reactor. (*Dagens Nyheter*, 1996 [our translation])

'One nuclear power station' would imply a potential of approximately 5 TWh (as compared to actual output of biopower of 9 TWh in 2006). Some enquiries convey a larger but still very limited potential. For instance SOU (2001, 77) suggests that the potential for biopower was about 12 TWh, whereas other studies suggest a potential that is at least 22–27 TWh, as reported by Jacobsson (2008).

An interesting aspect of the process of shaping expectations is in how opportunities are framed. For instance, in a Swedish government proposition it was argued that:

> The Government has over a number of years supported biomass based combined heat and power production (CHP). In CHP, the biomass is used with a very high efficiency. The available heat base sets, however, an upper limit to how much power can be generated with such a high efficiency. (Swedish Government 2002)

While technically correct, this citation portrays a picture of a potential that is tightly constrained by the district heating system and not the more appropriate image, in our view, of a vastly underutilised potential (Jacobsson 2008).

In sum, these examples indicate not only that what is considered relevant performance criteria may differ over time and vary between groups but also that besides apparent misinformation, there are many ways to frame and define a potential. This raises the question to what extent TA processes can be designed to increase transparency and reduce risk of misinformation. Despite efforts done in this direction, e.g. the International Organization for Standardization (ISO) 'Standards on Life Cycle Assessment Methodology', the inherent political nature of TA and the impossibility to create universally relevant, acceptable as well as stable system boundaries and performance criteria, means that TA is probably doomed to remain a contested, although necessary, activity.

A first challenge for policy makers is, therefore, to make room for and listen to a variety of voices, arguments and interpretations. A second challenge is to develop the competence and independence to critically assess various attempts to shape expectations and normative legitimacy by developing an own vision of a new technology. A developed vision that is ambitious enough may then underpin present and future policies (i.e. the institutional framework).[14]

Legitimation is, however, not only built on arguments about expected performance derived from assessments but also on the accumulation of actors in a TIS, the experience of and familiarity with a new technology as well as trust in various actors supporting it. As the German experience of wind and solar power demonstrate, institutions are shaped in a process of cumulative causation where entry of firms (and other organisations) and the formation and strengthening of advocacy coalitions are additional constituent structural parts (Jacobsson and Lauber 2006). Legitimacy of a new technology is, therefore, not only a prerequisite for the formation of a TIS but also a result of that very same process.

Clearly, changing the regulative framework constitutes a challenge not easily mastered by individual entrepreneurs in early phases of the development of a TIS. Instead, institutional entrepreneurship is expected to be pursued by coalitions of actors 'running in packs' (Aldrich and Fiol 1994; Van de Ven 1993), in which firms are concerned not only with '... their own immediate proprietary tasks and transaction modes but also with ... the overall social system' (Van de Ven 1993, 223). Yet, the very existence of such 'packs' presupposes a prior development of the TIS. In that development, 'knowledge development', although a prerequisite for system formation, can achieve little on its own. In particular, the formation of a market space that may allow for the building up of a critical mass of entrepreneurs who can form 'packs' engaging in institutional entrepreneurship is of critical importance (Kemp, Schot, and Hoogma 1998). Indeed, the clearest impact of the recent and very modest market formation programme for solar cells in Sweden (a few MW) has been the entry of firms and the formation of an industry association, which will act as a key node in an emerging advocacy coalition (Palmblad et al. 2006).

This space does not have to be large initially. The case of Germany is revealing in this respect. The stock of wind turbines in Germany in 1989 (end of formative phase) amounted to 221 machines with a capacity of 20 MW (Bergek and Jacobsson 2003). For solar power the stock in 1998 (end of formative phase) was 54 MW (Jacobsson et al. 2004). Similarly, there were only about 900 flexifuel vehicles in Sweden at the end of the formative phase in 2001 (Sandén and Jonasson 2005).

The market space has, however, to be large enough to sustain enough number of actors to make up a critical mass. In the Swedish wind power case, the first demonstration programme in the 1980s only allowed for 2 MW turbines to be built and the first commercial market consisted of isolated investments by farmers and economic associations. This was not enough to provide a market space to support early entrepreneurial initiatives. When investment subsidies were introduced in the early 1990s, the Swedish TIS was therefore weak and the market was handed over almost entirely to Danish firms. These did not have much interest in influencing the Swedish institutional framework to the benefit of further diffusion since there were other, rapidly expanding markets available (especially in Germany). Positive feedbacks failed, therefore, to materialise. This was, for instance, shown in the persistence of substantial weaknesses in municipal land-use planning and permits review processes. As a result of the stunted early development of TIS structure, there was no strong advocacy coalition and, thus, no institutional entrepreneurship to speak of to handle this issue. It was left largely to policy makers to achieve institutional alignment through various top down instruments (such as the appointment of areas of national interest for wind power and guidelines for local land-use planning). At least so far, these have not been successful (Bergek 2008).

Actor entry is, thus, obviously central to the process of acquiring legitimacy, which is sometimes forgotten when regulative change is viewed as an exogenous process of 'rational' choices based on 'rational' assessments. For instance, the inclusion of politically powerful agricultural industry organisations in the ethanol TIS has been of great importance for achieving regulative change (Sandén and Jonasson 2005). Moreover, for advocacy coalitions in the methane as well as ethanol systems, it has always been a prime target to enrol the Swedish car companies, not only to be able to materialise the technology (get vehicles) and learn from experimentation, but also and, perhaps more importantly, to make use of their political power. The launch of the Saab Biopower ethanol car in 2005, with a very visible advertising campaign, was a hallmark in the development of alternative fuels and signalled that the formative phase was definitely over.

Clearly, a bottom up evolutionary process to gain legitimacy can work if well orchestrated. Actor entry, growth of experience and normative and regulatory change may form virtuous circles enabling an expansion from small experiments to a system ready for take-off into a growth phase. A major policy issue is, therefore, to complement conventional science and technology policy measures with policies leading to the formation of early and protected market spaces. These are needed, not only to enable learning and to maintain a variety of voices and arguments, but also to enable experienced and trust based legitimacy as well as the build up of 'packs' of entrepreneurs and associated advocacy coalitions.

3.2. *'Development of positive external economies'*

The conventional view in management is that there is a competitive relationship between firms driving different designs within an emerging TIS (e.g. thin film *vs* crystalline silicon solar cells). In the literature, the notion of competing design is a central one and the formative phase is seen as a struggle for the selection of a dominant one (e.g. Utterback 1994). Hence, the advocates of a specific design are seen to compete for resources and markets in order to be associated with a new dominant design.

In the climate policy literature, the notion of competition is shifted to the relationship between different emerging systems (e.g. solar power *vs* wind power). The predominance of neoclassical economics has led to an emphasis on achieving 'cost efficient' solutions which leads to the choice of policies such as tradable green certificates where different emerging TIS are pitted against each others.

A diverging view is one where each emerging TIS centred on carbon neutral technologies, may contribute to destabilise the energy 'regime' (Raven 2005), by increasing both its own functionality and that of related TIS. As argued above, 'running in packs' is a strategy to increase functionality *within* a TIS. A similar notion of mutually beneficial co-evolution between several emerging TIS is based on relatedness that crosses apparent system boundaries.[15]

Marshall (1890) underlined the importance of economies that are external to the firm but internal to an industry (and location). Porter took that analysis a step further and incorporated also related industries into the realm in which external economies flow: 'External economies . . . often extend to related industries within the nation' (Porter 1990, 144) Related industries are '. . . those in which firms can coordinate or share activities in the value chain. . .' (Porter 1990, 105) In terms of our framework, relatedness is present if two or more emerging TIS share structural elements: actors and networks, technology or institutions.

Thus, we consider not only overlaps in terms knowledge spillovers (see, e.g. Eliasson 1997), but relatedness in terms of all structural components. To the extent that structural elements are shared, related TIS may mutually benefit from functions shaping these elements. The strengthening of one function in a TIS may, therefore stimulate the development of positive externalities that contribute to the build up of the structural components of another TIS.[16]

The socio-technical overlaps between different TIS blur the distinction between them. Indeed, the choice of system boundary in terms of level of aggregation, number and types of applications and value chains and geographical scope is somewhat arbitrary.[17] Ultimately, the choice depends on problem definition, i.e. what growth process that is to be studied. However, this does not imply that the choice of boundary is unimportant. If we choose a narrow system boundary, such as 'wheat ethanol production in Sweden' or 'wind power generation in Germany' we face the risk of missing important feedback mechanisms (treated as merely exogenous influence). However, applying a wide system boundary, e.g. including all technologies within the groups of renewable electricity production and renewable fuels respectively, or the even wider group of all renewable energy systems, implies that important feedback mechanisms may be missed due to lack of detail.

A possible solution is to apply narrower and wider system boundaries in parallel and that is exactly what we do when we point out the existence of positive externalities flowing between what we initially treated as separate TISs.

The history of alternative fuels in Sweden provides a clear example of structural overlaps and positive externalities. Even after three decades of development, variety is still increasing. In this process of generating and upholding variety, we can identify a number of instances where benefits 'spilled over' from one TIS to another. Indeed, we are inclined to say that a dominant feature is that the alternatives have complemented each other in the quest for system transformation.

For example, wheat ethanol and wood ethanol share the same downstream technology (vehicles that can run on ethanol). The short term option wheat ethanol has been legitimated as a bridging technology towards wood ethanol that is considered to have larger potential. This shared legitimacy was important to get tax exemptions, i.e. changed regulation, which generated market creation and actor entry into the wheat ethanol system. However, wood ethanol has benefited in terms of resource mobilisation (R&D investments) from the lobbying (legitimation) activities for ethanol by agricultural organisations. To broaden the picture of spillovers, ethanol fuels, in general, benefited from the development of a methanol TIS, in an earlier period (1974–1985). The development of knowledge of alcohol engines enabled entrepreneurial experimentation with ethanol and a decreased tax enabled market formation for ethanol buses in the 1990s. A third example is the development of gasification technology to produce methanol from bioenergy and coal in the 1970s and 1980s, as well as the establishment of actors carrying the competence, that are now, some 25 years later, key components in TISs centred on other types of fuels produced from gasified biomass (DME, FT-diesel, biogas and hydrogen). These actor networks were also key to legitimate some of these fuels, which resulted in a resource mobilisation (funding of demonstration plants) in recent years. To give a final example, organisations promoting environmental cars by procurement and labelling in Stockholm, Göteborg and Malmö, once developed to promote electric vehicles, created the first real markets for ethanol and biogas cars, i.e. shared actors resulted in market formation.

These examples may give the impression that the externalities can be predicted in a straightforward way. However, the complex pattern of interacting functions across narrowly defined TIS boundaries may create surprising results. The somewhat unique development of biogas as an alternative fuel in the transport sector in Sweden is an example of this. A 'search' process in the municipality of Linköping was directed towards methane as a result of the expected extension of the natural gas grid and a successful experimentation with methane (natural gas) buses in Malmö in the 1980s. In the meantime, an experiment with locally available biogas (also a methane rich gas) was initiated. When the extension of the natural gas pipeline was stopped in 1991, biogas use was expanded. This expansion was made possible by two parallel developments. First, a continued development of natural gas buses in Malmö and Göteborg created 'knowledge development' and 'materialisation' of methane bus technology. Second, a national demonstration programme for biofuels in heavy vehicles was launched to investigate (and stimulate) ethanol use. This demonstration programme surprisingly came to supply financial resources also to biogas and enabled market formation. In turn, this helped create the actors, networks, knowledge and legitimacy needed to establish a biogas transport fuel TIS with the momentum required to generate new resources and market shares.

In attempts to realise their potential, the growth of one TIS may, thus, pave the way for others. As is clearly seen in Germany, a legislative change promoting renewable energy technology (a feed-in law in 1991) that was initially driven by largely small scale hydropower advocates came in an unpredictable way, to benefit not only wind power in a massive way (Bergek and Jacobsson

2003), but also, with additional advocacy, solar power (Jacobsson and Lauber 2006). As Negro demonstrates, it also led to strategic investments by firms to develop a turnkey capability in the biogas field, which made possible the realisation of market opportunities created by the very same feed-in-law (Negro 2007). The initial efforts to align institutions to the need of small scale hydro power, thus, eventually came to benefit several other emerging TIS, all sharing a cognitive relatedness only.

The opportunity to share structural elements between what is normally seen as competing technologies, and competing TIS, open us for a further development of Van de Ven's (1993) notion of 'running in packs' as described above. Such packs, we would argue, may consist not only of firms sharing an interest in a specific design (e.g. CIGS solar cells) or a single TIS (solar cells) but also across a range of TIS. As underlined above, the relatedness may not be in the technical dimension at all but it may suffice to have cognitive relatedness. For instance, solar cells and biopower share nothing but the notion of 'renewable'. This is, however, a powerful relatedness that may be utilised to promote a vision of the future in which these two technologies jointly contribute to a plausible carbon neutral energy system (managing the politics of expectations).

Exploiting structural and functional overlaps clearly opens up for a more powerful 'bottom-up' process of institutional alignment and system growth than if each TIS is seen as self-contained. This has implications for both management and policy. For managers, it becomes essential to align the interests of advocates of several 'competing designs' and, in particular, to work *across* TIS by forming broader advocacy coalitions. Policy makers need to support such broader coalitions by selecting support schemes that do not pit various emerging TIS against each others in a misguided notion of efficiency that presupposes that the issue at stake is the selection of the most cost-efficient technology 'from the shelf'. Instead, they should favour regulatory frameworks that enable a number of TIS to go through a formative phase in parallel, drawing strength from structural and functional overlaps, thus 'filling the shelf' with a diverse set of technologies.[18]

4. Conclusions, implications for decision makers and an idea for further research

The purpose of this paper was to go some way towards contributing to an improved understanding of the key micro level processes in the formation phase of a new technological innovation system (TIS) and to identify associated policy and management challenges. Empirically, we have used our previous work on the development of renewable electricity technologies and transportation fuels in Sweden and Germany. We proposed an extension from a structural focus on dynamics to a combined structural–functional perspective, where a number of key processes – 'functions' – driving the development, diffusion and utilisation of new technology were identified. We discussed how these functions are shaped by both endogenous and exogenous factors and how they, in turn, shape the structural evolution of the TIS.

Whereas all these functions may guide decision makers seeking to engage in system building activities, we focused our analysis on two of the functions: 'legitimation' and 'development of positive externalities' where the angle to the latter was overlaps between different narrowly defined TISs. For the former, we pointed to the problematic role of technology assessment studies in 'the politics of expectations' and in shaping legitimacy. While necessary to reduce uncertainty, they also constitute a source of uncertainty and contain much opportunity to shape expectations in ways that are subtle and difficult to unmask. This is particularly problematic for immature technologies that face entrenched incumbents with a superior access to funding, media and politicians. Decision makers therefore need to not only preserve a variety of voices but also develop their own and independent competence to critically assess attempts to shape expectations and to form an own vision of the future of a specific technology.

The shaping of expectations is part and parcel of a bottom-up strategy of system building where 'packs of entrepreneurs' and others work to improve legitimacy, influence the direction of search of other firms, shape institutions and form markets. Yet, the existence of such 'packs' presupposes an initial market. It does not have to be large initially but is required for a process of cumulative causation to eventually unfold. Policy makers, therefore, have to form a regulatory framework that provides a smaller 'protected space' for 'packs' to emerge in, long before the technology is competitive in the market. Market formation is, thus, not only the effect of the development of new technologies but also a prerequisite for it.

To the extent that there are structural overlaps between different TISs, they may mutually benefit from functions strengthening these elements. In various ways, the achievement and growth of one TIS may pave the way for others – positive external economies may flow between different TISs. As shown in the case of biofuels in Sweden, positive interaction between TISs does not have to be in any way intended and part of strategy, but sometimes happens unexpectedly. However, exploiting overlaps opens up for a more powerful 'bottom-up' process of system growth than if each TIS is seen as self-contained. Managers and policy makers may enable this by going against conventional wisdom, forming broad based advocacy coalitions (management) and by fostering several TIS in parallel (policy).

The framework presented has proved its usefulness but additional work is required to develop it further. We will point to one line of direction for improvement. In Section 2, we argued that structural and functional dynamics were intertwined in the formative phase. For example, the function 'guide the direction of search' is closely linked to firm entry and 'legitimation' often involves shaping the institutional framework. Yet, our understanding of these feedback loops needs to be developed further. This applies, in particular, to the relation between functions and the formation of various types of networks. Entry of firms and other organisations upstream is a prerequisite for but does not automatically lead to formation of learning networks or political networks. Nor does market formation downstream automatically lead to the formation of user–supplier networks in which knowledge is exchanged beyond what is covered by a transaction. Hence, a limitation of our framework is that we do not fully capture the key processes that lead to network formation. In what follows, we will point at a line of argument which may deal with this weakness.

Arguably, networks are formed as and when the actors in the system realise the potential of the collective dimension of innovation systems. This dimension, as a whole, is captured in the seventh function, 'development of positive external economies'. Some of these are mediated through markets (e.g. pecuniary external economies stemming from a further division of labour) whereas others are connected to the functioning of networks, and therefore, flow only in as far as firms commit resources to investing in forming and maintaining networks. Hence, in order to interact, firms and other actors need to identify themselves as part of a system, see the common problems and opportunities they face and the value of collective action (Turner 2001). This was, for example, the case in the example of the Swedish solar cell TIS, as described in Section 2, where some firms saw the need for an organisation that could manage such action. In essence, therefore, network formation reflects the consciousness and practical realisation of parts of the collective dimension of the innovation and diffusion process. Without such a consciousness, user–supplier relationships will be arms-length, university–industry relationships may not develop and political networks will not be formed.

Acknowledgement

This paper was funded by the Swedish Energy Agency and carried out in the context of the project 'Policy Challenges and Options in the Transformation of the (Swedish) Electricity and Transport Sectors'.

Notes

1. The cases are documented in Bergek (2008), Bergek and Jacobsson (2003), Jacobsson (2008), Jacobsson , Sandén, and Bångens (2004), Palmblad et al. (2006), Sandén and Jonasson (2005).
2. Although the system concept may suggest collective and coordinated action, an innovation system is primarily an analytical construct. That we use the notion of an 'overall function' does not imply that all actors in a particular system exist for the purpose of serving that function or are directed by that function. Actors do not necessarily share the same goal, and even if they do, they do not have to be working together consciously towards it (although some may be). Indeed, conflicts and tensions are part and parcel of the dynamics of innovation systems. Nor does it imply that the system in focus has to exist in reality as a full-fledged one. As will be evident from this paper, it may be emerging with very weak interaction between components. Even in a more developed innovation system, moreover, the interaction between components may very well be unplanned and unintentional rather than deliberate.
3. The development of a technological niche in the literature on strategic niche management corresponds to an early part of the formative phase.
4. For instance, in the quite successful cases of solar cells and wind power in Germany, the formative phases ended in 1999 and 1991 respectively whereas their history can be traced back to the 1970s and before. See also Carlsson and Jacobsson (1997) and Geels (2002).
5. Often, the shift to a growth phase is triggered by external events.
6. For lists on positive feedback mechanisms from technology adoption in the energy sector and the solar cell industry, respectively, see Sandén and Azar (2005) and Sandén (2005). However, not only virtuous but also vicious circles are possible (Negro, Hekkert, and Smits 2007).
7. The 'transition' between an early formative phase and a phase of resilience roughly corresponds to that described by Geels (2002), in which a 'niche' emerges and grows within a (changing) regime or forms the basis of a new regime.
8. The whole notion of functions was developed to handle an integration of technology-specific and more general influencing factors: 'These factors may be fully technology specific, but may influence several technological systems simultaneously. Hence, they can be derived from a system perspective using different units of analysis: technology, industry, nation' (Johnson and Jacobsson 2001, 93).
9. The pattern of legitimation and development of positive externalities are likely to change as the systems grow and mature. However our scope in this paper does not extend beyond the formative phase where institutions are formed by legitimation and where systemic overlaps, or positive externalities, start to appear.
10. This refers primarily to normative and cognitive institutional alignment. Influencing the regulatory framework usually requires some legitimacy already to be in place.
11. Rightly so, strategic niche management emphasise this theme by identifying one of three key processes as 'voicing and shaping expectations' (Raven 2005). For the role of expectations in the case of gasified biomass in the Netherlands, see Negro, Hekkert, and Smits (2007).
12. The selection of timeframe is particularly important and is often done in an arbitrary or even seductive way and ultimate physical constraints are often confused with short or medium term technical, economic or political constraints. See Hillman and Sandén (2008).
13. This trauma has been created by a pressure to dismantle the nuclear power stations which run with a very low marginal cost and replacing that power production by new plants with a higher average cost.
14. Moderate expectations may keep some actors from entering the TIS. In the case of wind power, a national planning target of 10 TWh of wind power by 2015 was set by the government in 2002. Although this target is perceived to be reasonable by several types of actors, the corresponding county targets (which only amount to 4 TWh in total) seem to have been set on a too low level to facilitate wind power planning on regional and local levels (Bergek 2008). However, grand expectations could mobilise delegitimation efforts by threatened incumbents (Jacobsson and Bergek 2004).
15. For a related observation of symbiosis and predator–prey relationships between technologies besides pure competition, see Pistorius and Utterback (1997).

16. A structural overlap could, of course, also imply the risk that a bad reputation for one technology spills over and generates negative externalities for a related technology. See the example of nuclear power and the discussion on reputation spill-over in Sandén (2004).
17. We need to decide the level of aggregation in terms of technology (e.g. solar electricity, solar cells, thin film solar cells, CIGS thin-film solar cells, etc), the range of applications and value chains to include (e.g. ethanol production *vs* ethanol used as transportation fuel *vs* ethanol from forestry *vs* ethanol from forestry used as transportation fuel), and finally a geographical boundary (World, Europe, Sweden, county of Sweden) (cf. Bergek et al. 2008a).
18. This variety of course has the additional advantage of handling uncertainties from a societal level and building a response capacity. As and when a large scale deployment is deemed desirable, the capabilities and structures are there to respond, cutting lead time in the transformation of the energy sector.

References

Aldrich, H.E., and C.M. Fiol. 1994. Fools rush in? The institutional context of industry creation. *Academy of Management Review*, 19, no. 4: 645–70.

Alfsen, K., and G. Eskeland. 2007. A broader palette: the role of technology in climate policy. Report to the Expert Group for Environmental Studies 2007:1. Ministry of Finance, Stockholm.

Asheim, B.T., and A. Isaksen. 1997. Localisation, agglomeration and innovation: towards regional innovation systems in Norway? *European Planning Studies* 5, no. 3: 299–330.

Ashforth, B.E. and B.W. Gibbs. 1990. The double-edge of organizational legitimacy. *Organization Science* 1, no. 2: 177–94.

Bergek, A. 2002. Shaping and exploiting technological opportunities: the case of renewable energy technology in Sweden. Ph.D. thesis, Chalmers University of Technology, Göteborg.

———. 2008, under review. Levelling the playing field? The influence of national wind power instruments on local land-use planning practices in a Swedish county. *Energy Policy*.

Bergek, A., and S. Jacobsson. 2003. The emergence of a growth industry: a comparative analysis of the German, Dutch and Swedish wind turbine industries. In *Change, transformation and development*, eds. S. Metcalfe and U. Cantner, 197–227. Heidelberg: Physica-Verlag.

Bergek, A., S. Jacobsson, B. Carlsson, S. Lindmark, and A. Rickne. 2008a. Analyzing the functional dynamics of technological innovation systems: a scheme of analysis. *Research Policy* 37, no. 3: 407–29.

Bergek, A., S. Jacobsson, M. Hekkert, and K. Smith. 2008b, forthcoming. 'Functionality of innovation systems as a rationale and guide in innovation policy. In *Innovation policy, theory and practice. An international handbook*, eds. R. Smits, S. Kuhlmann and P. Shapira. Cheltenham: Elgar Publishers, in press.

Bijker, W.E. 1995. *Of bicycles, bakelites, and bulbs: toward a theory of sociotechnical change*. Cambridge, MA: MIT Press.

Carlsson, B., and S. Jacobsson. 1993. Technological systems and economic performance: the diffusion of factory automation in Sweden. In *Technology and the wealth of nations*, eds. D. Foray and C. Freeman, 77–94. London: Pinter.

———. 1997. In search of a useful technology policy – general lessons and key issues for policy makers. In *Technological systems and industrial dynamics*, ed. B. Carlsson, 299–315. Boston: Kluwer Press.

Carlsson, B., and R. Stankiewicz. 1991. On the nature, function, and composition of technological systems. *Journal of Evolutionary Economics* 1, no. 2: 93–118.

Carroll, G. 1997. Long-term evolutionary changes in organizational populations: theory, models and empirical findings in industrial demography. *Industrial and Corporate Change* 6: 119–43.

Dagens Nyheter. 1996. Ved kan inte ersätta kärnkraft. Biobränslen kan ersätta annan energi men bara marginellt bidra till elproduktion. *Dagens Nyheter*, 17 October.

Das, S.S., and A.H. Van de Ven. 2000. Competing with new product technologies: a process model of strategy. *Management Science* 46, no. 10: 1300–16.

David, P.A. 1988. Path dependence: putting the past into the future of economics. Institute for Mathematical Studies in the Social Sciences Technical Report 533. Stanford University.

Deeds, D.L., P.Y. Mang, and M.L. Frandsen. 2004. The influence of firms' and industries' legitimacy on the flow of capital into high-technology ventures. *Strategic Organization* 2, no. 1: 9–34.

Dosi, G. 1982. Technological paradigms and technological trajectories. *Research Policy* 11: 147–62.

Eliasson, G. 1997. General purpose technologies, industrial competence and economic growth. In *Technological systems and industrial dynamics*, ed. B. Carlsson, 201–53. Boston: Kluwer Press.

Fransman, M. 1990. *The market and beyond*. Cambridge: Cambridge University Press.
Galli, R., and M. Teubal. 1997. Paradigmatic shifts in national innovation systems. In *Systems of innovation*, ed. C. Edquist, 342–70. London: Pinter.
Geels, F., and B. Raven. 2006. Non-linearity and expectations in niche-development trajectories: ups and downs in Dutch biogas development (1973–2003). *Technology Analysis & Strategic Management* 18, no. 3: 375–92.
Geels, F.W. 2002. Technological transitions as evolutionary reconfiguration processes: a multi-level perspective and a case study. *Research Policy* 31, no. 8–9: 1257–74.
———. 2004. From sectoral systems of innovation to socio-technical systems: insights about dynamics and change from sociology and institutional theory. *Research Policy* 33, no. 6–7: 897–920.
Hellsmark, H. 2004. Incumbents versus new entrants in the exploitation of renewable energy technologies. mimeo, Department of Industrial Dynamics, Chalmers University of Technology, Göteborg.
Hillman, K.M., and B.A. Sandén. 2008. Time and scale in life cycle assessment: the case of fuel choice in the transport sector. *International Journal of Alternative Propulsion* 2, no. 1: 1–12.
Hughes, T.P. 1983. *Networks of power – electrification in Western society 1880–1930*. Baltimore, MD: The Johns Hopkins University Press.
———. 1987. The evolution of large technological systems. In *The social construction of technological systems: new directions in the sociology and history of technology*, eds. T.P Hughes, W. E. Bijker and T. Pinch. Cambridge, MA: MIT Press.
Jacobsson, S. 2008. The emergence and growth of a 'biopower' innovation system in Sweden. *Energy Policy* 36, no. 4: 1491–508.
Jacobsson, S., and A. Bergek. 2004. Transforming the energy sector: the evolution of technological systems in renewable energy technology. *Industrial and Corporate Change* 13, no. 5: 815–49.
Jacobsson, S., and A. Johnson. 2000. The diffusion of renewable energy technology: an analytical framework and key issues for research. *Energy Policy* 28: 625–40.
———. 2001. Inducement and blocking mechanisms in the development of a new industry: the case of renewable energy technology in Sweden. In *Technology and the market: demand, users and innovation*, eds. R. Coombs, K. Green, V. Walsh and A. Richards. Cheltenham: Edward Elgar.
Jacobsson, S., and V. Lauber. 2006. The politics and policy of energy system transformation – explaining the German diffusion of renewable energy technology. *Energy Policy* 34, no. 3: 256–76.
Jacobsson, S., B.A. Sandén, and L. Bångens. 2004. Transforming the energy system – the evolution of the German technological system for solar cells. *Technology Analysis & Strategic Management* 16, no. 1: 3–30.
Johnson, A. 1998. Functions in innovation system approaches. mimeo, Department of Industrial Dynamics, Chalmers University of Technology, Göteborg.
———. 2001. Functions in innovation system approaches. Electronic paper at the Nelson and Winter Conference, Aalborg, Denmark, June 2001.
Kaiser, A. 1992. Redirecting power: Swedish nuclear power policies in historical perspective. *Annual Review of Energy Environment* 17: 437–62.
Kemp, R., J. Schot, and R. Hoogma. 1988. Regime shifts to sustainability through processes of niche formation: the approach of strategic niche management. *Technology Analysis & Strategic Management* 10, no. 2: 175–95.
Layton, E.T. 1974. Technology as knowledge. *Technology and Culture* 15, no. 1: 31–41.
Marshall, A. 1890. *Principles of economics*, 8th edn. London: Macmillan and Co.
Myrdal, G. 1957. *Economic theory and underdeveloped regions*. London: Ducksworth.
Negro, S.O. 2007. *Dynamics of technological innovation systems: the case of biomass energy*. Utrecht, Copernicus Institute for Sustainable Development and Innovation.
Negro, S.O., M.P. Hekkert, and R.E. Smits. 2007. Explaining the failure of the Dutch innovation system for biomass digestion – a functional analysis. *Energy Policy* 35, no. 2: 925–38.
Oliver, C. 1991. Strategic responses to institutional processes. *Academy of Management Review* 16, no. 1: 145–79.
Palmblad, L., S. Jacobsson, M. Hall, and B.A. Sandén. 2006. Dynamics of the Swedish PV innovation system – the impact of a recent market formation programme. Paper presented at the 21st European Photovoltaic Solar Energy Conference, Dresden, 4–8 September 2006.
Pistorius, C.W.I., and J. M. Utterback. 1997. Multi-mode interaction among technologies. *Research Policy* 26: 67–84.
Porter, M. 1990. *The competitive advantage of nations*. London: McMillan.
———. 1998. Clusters and competition. New agendas for companies, governments and institutions. In *On competition*, ed. M. Porter, 197–287. Boston, MA: Harvard Business Review Press.
Rao, H. 2004. Institutional activism in the early American automobile industry. *Journal of Business Venturing* 19: 359–84.
Raven, R. 2005. Strategic niche management for biomass. PhD thesis, Centre for Innovation Studies, Eindhoven University.

Rosenbloom, R.S., and C.M. Christensen. 1994. Technological discontinuities, organizational capabilities, and strategic commitments. *Industrial and Corporate Change* 3, no. 5: 655–85.
Sabatier, P.A. 1998. The advocacy coalition framework: revisions and relevance for Europe. *Journal of European Public Policy* 5: 98–130.
Sandén, B.A. 2005. The economic and institutional rationale of PV subsidies. *Solar Energy* 78: 137–46.
———. 2006. Technology path assessment for sustainable technology development. *Innovation: Management, Policy and Practice* 6: 316–30.
Sandén, B.A., and C. Azar. 2005. Near-term technology policies for long-term climate targets: economy wide versus technology specific approaches. *Energy Policy* 33: 1557–76.
Sandén, B.A., and K.M. Jonasson. 2005. Variety creation, growth and selection dynamics in the early phases of a technological transition: the development of alternative transport fuels in Sweden 1974–2004. ESA report 2005:13, Chalmers University of Technology, Göteborg.
Simon, H.A. 1955. A behavioral model of rational choice. *The Quarterly Journal of Economics* 69, no. 1: 99–118.
SOU (Statens Offentliga Utredningar). 2001. *Handel med certifikat. Ett nytt sätt att främja el från förnybara energikällor* (Stockholm, Statens Offentliga Utredningar).
Stern, N. 2006. *The economics of climate change – the Stern review*. Cambridge: Cambridge University Press.
Suchman, M.C. 1995. Managing legitimacy: strategic and institutional approaches. *Academy of Management Review* 20, no. 3: 571–610.
Swedish Government. 2002. Prop. 2001/02:143: Samverkan för en trygg, effektiv och miljövänlig energiförsörjning (Collaboration for a safe, efficient and environment-friendly energy supply), Government bill, March 2002.
Teppo, T. 2006. Financing clean energy market creation – clean energy ventures, venture capitalists and other investors. Doctoral Dissertation Series 2006/1, Helsinki University of Technology.
Turner, R.C. 2001. *A framework for cluster-based economic development policies*. New York: The Nelson A. Rockefeller Institute of Government.
Unruh, G.C. 2000. Understanding carbon lock-in. *Energy Policy* 28, no. 12: 817–30.
Utterback, J.M. 1994. *Mastering the dynamics of innovation*. Boston: Harward Business School Press.
Van de Ven, A., and R. Garud. 1989. A framework for understanding the emergence of new industries, research on technological innovation. *Management and Society* 4: 195–225.
Van de Ven, A.H. 1993. The development of an infrastructure for entrepreneurship. *Journal of Business Venturing* 8: 211–30.
Van Lente, H. 1993. Promising technology. In *The dynamics of expectations in technological development*, ed. A. Rip. Twente: Universiteit Twente.
Wenk, E., and T.J. Kuehn. 1977. Interinstitutional networks in technological delivery systems. In *Science and technology policy*, ed. J. Haberer, 153–75. Lexington: Lexington Books.
Zimmerman, M.A., and G.J.F. Zeitz. 2002. Beyond survival: achieving new venture growth by building legitimacy. *Academy of Management Review* 27, no. 3: 414–31.
Zucker, L.G. 1987. Institutional theories of organization. *Annual Review of Sociology* 13: 443–64.

Cumulative causation in biofuels development: a critical comparison of the Netherlands and Sweden

Karl M. Hillman, Roald A.A. Suurs, Marko P. Hekkert and Björn A. Sandén

Supporting the development and diffusion of sustainable innovations has become a dominant topic on the political agenda of many countries. However, this has proven to be a difficult task. To increase insight in such processes, this paper takes biofuel technologies in the mobility sector as the topic of a comparative case study. Various national governments have supported innovation trajectories around biofuels. We analyse, assess and compare two such trajectories as they have developed so far: one in the Netherlands and one in Sweden. A Technological Innovation System (TIS) approach is applied. A TIS is constituted by actors, networks and institutions, that are to be gradually constructed around a technology. We analyse whether governments and entrepreneurs have succeeded in developing seven key processes, or system functions, necessary for the development and diffusion of biofuel technologies. By analysing the build-up of system functions over time we identify virtuous and vicious forms of cumulative causation. The Dutch and Swedish TISs for biofuels are followed from 1990 to 2005. Our comparison shows that, due to a variety in function fulfilment and the emergence of cumulative causation, the Swedish TIS has reached a market expansion and broad social implementation of biofuels, whereas the Dutch TIS has established considerably less.

1. Introduction

The problem of climate change – as a result of greenhouse gas emissions – is today (2008) seen as one of the most important environmental issues. A major culprit is the transport sector, which has been growing for several decades (IEA 2004). Many scenarios assume that this sector will continue to increase its share of world fuel use for many decades (EIA 2003; IEA 2006). The emissions from the transport sector can be mitigated by a reduction in transport activity and more efficient vehicles, but there is also a need for new sources of energy that – on a net balance – do not (or barely) give rise to greenhouse gas emissions. Thus, strong arguments are made for public support of renewables (Sandén and Azar 2005), and there is an increasing political willingness to particularly support biomass-based transportation fuels: i.e. biofuels (IEA 2004). In the European Union (EU), this has been acknowledged in a policy directive, which presses the national governments to enforce the introduction of biofuels (EU 2003).

Still, the task of supporting renewables is not trivial. The emergent technology will have to compete within the environment of a techno-institutional complex: a set of interlinked 'large technological systems and the public and private institutions that govern their diffusion and use' (Unruh 2000, 825–826). Over the past decades, this complex has evolved into an energy and transport system that continuously provides positive feedback – in the form of scale economies, cumulative learning, network externalities and the emergence of habitual routines – for the further development of fossil based technologies (Unruh 2002). Hence, diffusion of emerging renewable technologies is hampered by the momentum of the energy and transport systems.

Nevertheless, change is possible. Studies by Suurs and Hekkert (2008) and Sandén and Jonasson (2005) have investigated the recent history of biofuels in the Netherlands and Sweden, respectively. Both studies show how the technology passes through a formative stage characterised by weak institutions, little market demand and multiple technology designs. The returns on (private) investments are highly uncertain and only to be expected in the long run – developments largely depend on public support.[2] However, in Sweden, entrepreneurial efforts and government policies linked up in a process of cumulative causation (as a result of positive feedback) (Myrdal 1957) putting significant pressure on the transport system to adopt biofuels, with the result that utilisation increased, starting from the mid 1990s, to more than 2% of the total transport fuel supply in 2004 (Sandén and Jonasson 2005). In the Netherlands, the government also supported biofuels, but market diffusion has – at the time of writing – barely taken off (Suurs and Hekkert 2005, 2008). How can this difference be explained and what lessons can be learned?

The purpose of this paper is to point out the drivers and barriers for cumulative causation in the development of biofuels in both the Netherlands and Sweden (1990–2005). Questions to be answered are:

(1) What patterns of cumulative causation occurred in the two countries?
(2) What caused the patterns to emerge or collapse?
(3) What lessons may we draw for policy makers and entrepreneurs?

One important point to make is that in this paper we are interested in dynamics and growth processes in a formative stage. We do not attempt to say anything about the desirability of introducing various types of biofuels. Obviously, the desirability as perceived by the actors in the system does play a role in our analysis as these positive and negative expectations are crucial to understanding innovation system dynamics. We also admit that in later growth stages, beyond the scope of this study, limitations of various kinds may very well hamper a sustained growth (as a result of negative feedback).

Studying the formative stage of technological trajectories demands an analytical framework that addresses not just the technical and economical dimensions of technological trajectories but also the broader processes of societal embedment (Bijker, Hughes, and Pinch 1987). Moreover, because the formative stage is characterised by rapid changes in the system surrounding and supporting the technology, the analysis needs to pay close attention to the dynamics of change within this system (Hekkert et al. 2007; Jacobsson and Bergek 2004). There are many partly related streams of literature that in different ways take these dimensions into account (Bijker 1987; Van de Ven 1993). In this study we choose to adopt the framework of Technological Innovation Systems (TISs). The TIS approach stresses the idea that innovation is a collective and context specific process, wherein governments have a particular role of facilitating the introduction of emerging technologies (Jacobsson and Johnson 2000). The recent TIS literature that we build upon explicitly tries to conceptualise the dynamics related to a technological field (Hekkert et al. 2007). Much attention has been given to systematic operationalisation of the concepts used (Hekkert et al. 2007), and the system metaphor opens for assessment of system performance and effectiveness of policy and entrepreneurial intervention.

The paper is structured as follows: In Section 2, an overview of theoretical concepts is presented, together with a methodology for operationalising the concept of cumulative causation. Section 3 presents a background to biofuels development, and Sections 4 and 5 provide the analysis of the Dutch and Swedish TISs related to biofuels, here called Biofuels Innovation Systems (BISs). In Section 6, a comparison of the observed dynamics is given, resulting in some strategic insights. Conclusions are formulated in Section 7.

2. Functions of innovation systems

Various innovation system approaches exist: from nationally oriented approaches to those with a focus on regions or industrial sectors. In our case we apply the framework of Technological Innovation Systems (TISs). In terms of its boundaries a TIS can be regarded as 'cross-cutting' geographical and sectoral domains; the principle starting point is a technology field. As an additional delineation we focus on the national TISs of the Netherlands and of Sweden. In the greater part of the literature, the TIS is defined by the actors (organisations), networks of actors and institutions (regulations, norms, cultures) involved in the generation, diffusion, and utilisation of a technology.[3]

The TIS framework serves as a heuristic to point out the structural elements that are relevant for a strategic intervention in technology development (Jacobsson and Bergek 2004).[4] However, most innovation system approaches have focused on just this and, as a result, have neglected the development process itself. See, for instance, how Freeman contrasts various National Innovation Systems – Japan *vs* USSR in the 1970s, East Asia *vs* Latin America in the 1980s – in terms of R&D intensity, educational system, industrial structure, and many other factors (Freeman 1995). Even if processes are being measured, such as yearly budgetary flows or patent outputs, then still no attention is paid to the interdependence of such processes in time.[5] Important contributions to the sectoral innovation systems approach also show a primary focus on the state of structures: technological regimes rather than historical mechanisms (Breschi, Malerba, and Orsenigo 2000). The TIS approach has been recognised by Carlsson et al. (2002) to be the most dynamic member of innovation systems approaches and many scholars have used the TIS framework to conduct longitudinal studies that indicate the key historical drivers of a technological field.

A focus on static structure has some disadvantages: (1) as structural factors have been historically shaped, they will differ so much between contexts that a comparison between TISs is hard to make, and (2) a static structural approach tends to disregard important explanatory factors behind cumulative causation dynamics (Hekkert et al. 2007). To overcome these shortcomings, recent studies have suggested augmenting the analysis of structures with an analysis of key processes, or system functions (Carlsson et al. 2002; Bergek 2002; Jacobsson and Bergek 2004; Bergek et al. 2008; Hekkert et al. 2007; Edquist 2005). The literature presents different lists of such functions. In this paper we will use the seven functions listed in Table 1 (for an overview, see Bergek 2002; Hekkert et al. 2007).

Table 1. Functions of technological innovation systems.

F1. Entrepreneurial activities	At the core of any innovation system are the entrepreneurs. These risk takers perform the innovative commercial experiments, seeing and exploiting business opportunities.
F2. Knowledge development	Technology research and development are prerequisites for innovations, creating variety in technological options. R&D activities are often performed by researchers, but contributions from other actors are also possible.
F3. Knowledge diffusion	The typical organisational structure of an emergent innovation system is the knowledge network, primarily facilitating information exchange.
F4. Guidance of the search	Often within a transition trajectory, various technological options exist. This function represents the selection process that is necessary to facilitate a convergence in development, involving for example policy targets, outcomes of technical or economical studies and expectations about technological options.
F5. Market formation	New technologies often cannot outperform established ones. In order to stimulate innovation it is necessary to facilitate the creation of (niche) markets, where new technologies have a possibility to grow.
F6. Resource mobilisation	Material and human factors are necessary inputs for all innovation system developments, and can be enacted through e.g. investments by venture capitalists or through governmental support programmes.
F7. Support from advocacy coalitions	The emergence of a new technology often leads to resistance from established actors. In order for an innovation system to develop actors need to raise a political lobby that counteracts this inertia, and supports the new technology.

These functions are defined as being categories of 'actions taken by actors in the system' (events).[6] Since the system functions are defined on a more abstract level than the actions themselves, they are applicable to a wide variety of empirical observations. As a result, it becomes possible to compare different TISs in a systematic way. The system functions can be used to explain the endogenous dynamics of the system. However, any TIS is also embedded in a wider environment that influences these dynamics. This exogenous impact consists of factors beyond the direct influence of the national TIS, for instance oil price shocks and international policy.[7]

At the heart of the TIS approach is the idea that system functions can interact with each other (Bergek et al. 2008). For instance, the successful realisation of an important research project (knowledge development), may result in a rise of expectations among policy makers (guidance of the search), which may subsequently trigger the start-up of a subsidy programme (resource mobilisation) to support even more research projects (knowledge development). This example illustrates how complementary effects can result in the unfolding of a pattern of cumulative causation (Myrdal 1957).[8] Events drive other events that in turn set in motion new events. As the sequence continues, it can branch out, eventually resulting in a richer realisation of system functions. Ideally, these dynamics may be cyclic, creating a loop of positive feedback, which may trigger a take-off from a formative to a growth stage of the TIS. However, positive feedback may also be destructive, yielding a regression of the TIS. In the first case we speak of virtuous dynamics (a build-up sequence) and in the second case of vicious dynamics (a break-down sequence).[9]

The empirical basis for the analysis of the TIS dynamics is a narrative consisting of sequences of key events.[10] Each event is mapped on a system function; this way the event serves as the empirical counterpart of a system function (Table 2). Patterns of events in the narrative can provide insight in the emergence of cumulative causation. Identification of patterns is based on

Table 2. Event types as indicators of innovation system functions; some events may contribute negatively to a system function (for instance expectations or lobbies).

System function	Event types
F1: Entrepreneurial activities	Projects with a commercial aim, demonstrations, portfolio expansions
F2: Knowledge development	Studies, lab trials, pilots, research programmes
F3: Knowledge diffusion	Conferences, workshops, alliances between actors
F4: Guidance of the search	Expectations, promises, policy targets, standards, research outcomes
F5: Market formation	Market regulations, tax exemptions, events constituting niche markets
F6: Resource mobilisation	Subsidy programmes
F7: Support from advocacy coalitions	Lobbies, advice

qualitative techniques presented in Poole et al. (2000). Here our focus is on interaction patterns, i.e. sequences of events. By relating event sequences to the abstract level of system functions, we can point out how system functions reinforce (or antagonise) each other through time. If this carries on for some time, this indicates a process of cumulative causation.[11]

The narratives are organised chronologically in episodes, where all episodes are characterised by different types of interaction patterns. When an event contributes to a system function, this is indicated within square brackets and numbering following Table 1 [F*x*]. A negative contribution is indicated by a minus sign [–F*x*]. Common background factors are introduced in the following section; exogenous factors that are case specific are only mentioned in the separate narratives.

3. Background to biofuels development

We identified five exogenous factors that drive or hamper the endogenous developments in both our cases: (1) oil supply, (2) local air and water quality, (3) EU agricultural policy, (4) climate change, and (5) technology features. These factors are of both international and national origin, but are common for both countries and are generally unaffected by the specific Biofuels Innovation Systems (BISs). As will become clear from our analysis, their influence on BIS dynamics differed between the two studied systems.

The first factor is the insecurity of oil supply, mainly represented by oil price shocks. The oil crises in the 1970s made the industrialised world aware of the vulnerability connected to oil dependence. Alternative energy sources were sought and increasingly biofuels technologies were being developed to utilise them. However, from the 1980s on, the oil price dropped and ceased to be a driver of change. Cheap fossil fuels then became an economic barrier for all renewables development. This only changed towards the end of the studied period, as the oil price began to increase.

The second factor concerns worries about local air and water quality. The issue arose on the political agenda throughout Europe from the 1980s on. In the transport sector this resulted in incremental innovations such as the catalytic converter and cleaner diesel engines. It also became an argument for implementing biofuels.

Third is EU agricultural policy. The issue at hand is a restructuring of the declining agricultural sector, reducing the agricultural surplus and exploiting new markets. From the 1990s, farmers received subsidies for keeping fallow grounds and for the cultivation of non-food crops. The climax was the 'Biofuels Directive' of 2003, in which targets were set for the member states regarding the introduction of biofuels.[12] The directive was increasingly legitimated by referring to energy security and climate change.

Fourth, the debate on climate change especially gained attention in the 1990s. Because of the possible reduction of CO_2 emissions, biofuels were seen as an interesting option. A milestone was the signing of the Kyoto treaty by the EU member states in 1998. The European Commission recognised the utilisation of biomass resources as having a great potential to reduce CO_2 emissions (EU 1997).

Finally, in addition to these factors, different technologies emerged of which the inherent features were affecting the events in our two cases. European policy has particularly supported biofuels based on agricultural products. This involves a rather conventional set of conversion technologies with limited performance in terms of costs and CO_2 reduction. An alternative exists: a second technology group with potentially lower costs and higher CO_2 reductions (Hamelinck and Faaij 2006). The differences between this first generation (1G) and second generation (2G) of biofuels have played a central role.[13]

In the following we will reconstruct the development of the Swedish and the Dutch BISs. For each country we will focus on that part of the technological domain where the most activities have taken place; 1G/2G bio-diesel in the Netherlands, and 1G/2G bio-ethanol in Sweden.[14]

4. The Dutch Biofuels Innovation System: the case of bio-diesel

Developments of biofuels for transport started around 1990. The case of diesel substitutes serves to illuminate the characteristic features of this development. Four episodes are distinguished.

4.1. *European drivers of change*

Developments of transport fuels started when, from 1990 on, small entrepreneurial experiments with the adoption of 1G biofuels were initiated [F1]. The projects received policy support and financing, partly from the EU (exogenous). Trials were performed [F2] and turned out to be technically successful [F4], but the economic returns were disappointing [–F4]. Regional governments played a guiding role during this early start [F4]. However, the national government stood back, as it doubted the importance of the small projects [–F4]. Moreover, a couple of influential assessment studies [F2] underlined the shortcomings of the agriculture-based 1G biofuels [–F4]. The argument was that they were expensive and offered no long term potential (*Algemeen Dagblad* 1992; *NRC Handelsblad* 1992).

Up to 1995 not much happened. In this first episode, developments were mainly driven by the EU policy which was being prepared (exogenous). The system functions were just beginning to take shape.

4.2. *The shaping of a niche*

The experiments with the adoption of agriculture-based fuels had limited success. A major drawback was the extreme scarcity of land, which pushed up the production costs of the energy crops (Landbouw-Economisch Instituut 2005). In contrast, another exogenous factor triggered a wave of new experiments, namely the increased stringency of regulation with respect to local water quality standards. Biofuels became a commercially feasible option for the utilisation in ships, as they are biodegradable and therefore less polluting.

A small niche market emerged [F5]. Two (private) boating companies in the rural province of Friesland grasped the opportunity and initiated projects with 1G bio-diesel [F1]. The companies started adoption trials [F2]; as they did this, they turned to government authorities, on the provincial and on the national level, to lobby for a tax exemption [F7]. The province – and the district board of agriculture – supported the idea and started to advocate biofuels as well [F7]. They were successful; a first tax exemption – for two years – was provided [F5] (*Het Financieele Dagblad* 1995).

The adoption experiment [F1] resulted in knowledge creation [F2] and most importantly it served as an example to others in the field [F4]. The province decided to adopt bio-diesel itself as well, for its fleet of service boats [F1]. Multiple other boating projects were started [F1]. Again tax exemptions were demanded [F7], and were in turn issued by the national government [F5]. The 1G technologies gained more and more attention [F4]. Still, the experiments were only partly successful; technically everything seemed to work, but the boats caused a smell and there was a negative economic pay-off [–F4]. Nevertheless some of the experiments continued till 2002 [F1].

The entrepreneurial (boating) experiments contributed to a first wave of cumulative causation, driven by advocacy coalitions of entrepreneurs and local governments and their consistent lobby for resources. This was exogenously conditioned by local water quality regulations.

Meanwhile, developments around 1G biofuels were heavily scrutinised by scientists, engineers, and most notably, by the environmentalist movement [–F4, –F7]. The national government remained divided, and as a result, entrepreneurs who performed adoption experiments [F1, F2] lacked the necessary support. Tax exemptions [F5] were incidentally issued, though for a short time period. No policy strategies were formulated, and hence (potential) bio-diesel entrepreneurs remained in the dark with respect to the future institutional support [–F4].

From 1997 the criticism on biofuels – aimed at 1G technologies – turned into a more positive movement as new options gained attention. From then on, the 2G technologies – which were being developed in the laboratories [F2] – became part of the public discussion [F4]. This was mainly the work of entrepreneurs who started attempts to bring 2G biofuels to the market [F1]. The Dutch government – mainly the ministry of VROM (the Ministry of Housing, Spatial planning and the Environment) – considered acquiring a stake in this promising emerging technology [F4].[15]

So 2G biofuels were starting to play a role during this episode, although this was largely limited to knowledge development. A downside was that now the experiments with 1G biofuels were shaded by the promises of 2G biofuels. No new entrepreneurial experiments were initiated.

4.3. *The GAVE programme initiates R&D*

In 1998 the climate issue became an important exogenous driver. As a response to the Kyoto protocol, the GAVE (Gasvormige en Vloeibare Energiedragers) platform is initiated in 1999 – through Novem.[16] The programme was the first programmatic government attempt to support biofuels [F4]. The transport sector was originally not its specific orientation, but transportation fuels were quickly established as a priority.[17] GAVE's first move – during 1999 – was to authorise a number of assessment studies [F4, F6] (GAVE 1999; KEMA Consultancy 2000). The studies should clear up the controversy around various biofuel options. The result was a shortlist of fuel chains to be analysed in more detail, based mainly on energy balances and cost figures [F2]. GAVE's advice was to exclusively support the projects where a CO_2 reduction of 80% was guaranteed possible [F4].[18] All 1G options were then (*de facto*) excluded from further assessments [F4] (GAVE 2003). It is within this context that the term 2G biofuel was actually introduced to distinguish the contested agricultural biofuels from technologically advanced options.[19]

GAVE provided €2 million [F6] to stimulate the formation of networks [F3] and to support assessment research [F2]. This should eventually lead to a demonstration for which a budget of €5 million was reserved [F6]. The emphasis was on innovation in fuel production, where cooperation with downstream industrial parties and private investors was encouraged [F3]. The projects that started [F1] were all exclusively oriented towards 2G options.

The projects were setup by two alliances – the Shell–ECN network[20] and the TNO network[21] – and other actors, among which were investors [F6], a car company and many others [F1, F3] (GAVE 2002). The projects were successful [F4], particularly with respect to solving important technical bottlenecks [F2] such as gas cleaning [F2] (Boerrigter, Den Uil and Calis 2002). The programme served as a catalyst that bundled and guided R&D projects that until then had been going on in relative isolation [F3, F4]. As a consequence multiple entrepreneurs [F1] started new biofuels projects during this episode, even outside GAVE.

The purpose of GAVE was to realise a demonstration plant. By the end of 2002 it turned out that the two networks were both viable candidates. Unfortunately both parties decided to decline [−F1]. As a result, the programme stopped [−F6] and failed to deliver. The main reason was that the building of a commercial plant would cost far more than €10 million; with the absence of a flanking market creation policy, this did not make an economic business case.

In this episode, a pattern of cumulative causation pivoted around entrepreneurial experiments related to 2G technology, but especially around guidance, knowledge diffusion and resource mobilisation delivered by GAVE. This has resulted in important successes. Knowledge has been developed and the urgency of developing sustainable transportation fuels was beyond discussion, but still there was no commercial perspective for biofuels. Besides the deficiencies of the national BIS, the slow progress with 2G biofuels can to some extent be seen as an exogenous factor, i.e. the requirements for advanced technology.

4.4. *A market driver*

Things changed in 2003 as the EU issued the Biofuels Directive (EU 2003). This exogenous factor had drastic consequences, as in contrast to the Dutch government, the EU was largely oriented towards 1G biofuels. With the subsidy programme terminated and with the new task of translating the EU directive to national policy, the national government reoriented its policy (Stromen 2003).

From 2003 on, GAVE was given a new priority task [F4]: the development of a generic market for biofuels. The 1G technologies were increasingly perceived as bridges towards 2G fuel implementation.[22] The 1G bio-diesel options had so far been excluded from support by GAVE [–F4], with reference to arguments such as high costs and weak environmental performance, implying that market-oriented projects had been subdued. Up to 2002, not a single drop of bio-diesel (1G or 2G) had been produced by Dutch companies.[23] By 2002–2003 this changed, as many regional entrepreneurs executed plans for the construction of small factories [F1]. The main driver was (again) the EU directive (exogenous).

These were the first commercial experiments that targeted the supply side of the biofuels chain [F1]. The projects were supported by a large number of actors; among them farmers, farmers' associations and local government authorities [F3]. Many of these actors were made shareholders [F6]. Also, biofuels were promoted to potential users [F4]. For these projects to work out financially, tax exemptions were requested [F7], and issued on a project basis [F5] (*Bizz* 2002a,b). By 2005, the first (1G) bio-diesel plant was built. This successful outcome [F4] triggered a pattern of cumulative causation, as from 2002 numerous projects [F1] started all over the country, especially in rural areas. The later developments were additionally driven by the

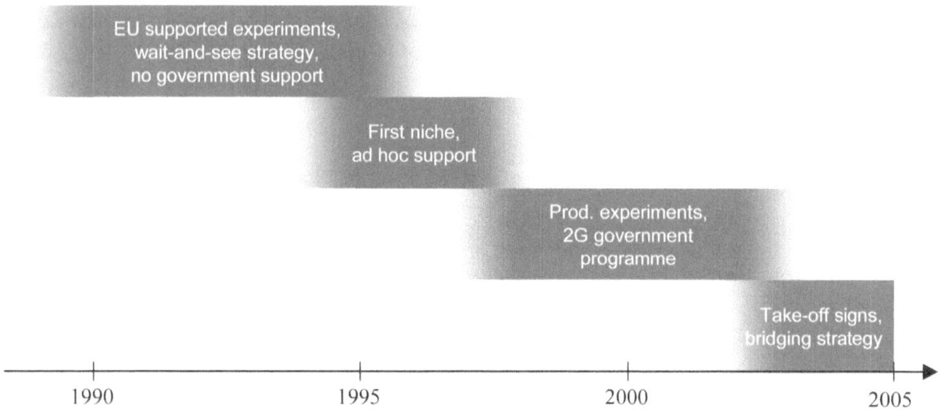

Figure 1. Four episodes of Dutch BIS development.

advocacy and guidance from the Energy Valley cluster [F7, F4], a regional network of public–private partnerships. The first plant became an often used example [F4] (*Dagblad van het Noorden* 2004, 2005; *Leeuwarder Courant* 2005; *Provinciale Zeeuwse Courant* 2004). In 2004 numerous municipalities started to adopt bio-diesel for their car fleets [F1] (Stromen 2004; Rotterdams Dagblad 2004).

In this period, multiple functions were fulfilled – bottom–up – by various collaborating actors. The promise of the (exogenous) EU directive had apparently been more important in fostering this interaction pattern than the ad-hoc national support. Remarkably, it was again regional authorities and entrepreneurs that took the BIS forward. The perspective of a market, offered through the EU directive, played a pivotal role. Also national policy makers realised that their scheme towards demonstrating fuel chains had failed because there was no market. In accordance with this, GAVE changed its role from an R&D catalyser to a facilitator of networks and markets. In the process, GAVE influenced guidance by stressing the concept of bridging technologies – from 1G to 2G. In summer 2004, the government released a white paper on traffic emissions. The document contained a section on generic measures to be taken for the implementation of biofuels. The 1G fuels were now explicitly regarded as a bridging option. Whether this will turn out to be a watershed event, leading to a take-off remains to be seen. See Figure 1 for an overview of the four episodes of Dutch BIS development.

5. The Swedish Biofuels Innovation System: the case of bio-ethanol

The development of alternative fuels in Sweden began in the 1970s with a large focus on methanol produced from different resources. This was a response to drastic increases in the oil price, but the interest faded out as the price went down in the early 1980s. At this time, the first activities related to bio-ethanol surfaced. Through the years, this development would come to revolve around four parallel tracks that were more or less intertwined and in different ways contributed to the build-up of the BIS. These were the bus niche, the production of wheat ethanol (1G), the market for flexifuel cars and wood ethanol research (2G). The Foundation for Swedish Ethanol Development (SSEU) was the central player in two of them, while partly influencing the other two. It was formed in 1983 by private and municipal companies around a plant processing by-product ethanol from the pulp and paper industry in Örnsköldsvik. After 1990, bio-ethanol became the dominating biofuel and it serves as an illustrative example of the development of biofuels in Sweden, during six episodes.

5.1. *1980s pre-developments*

After some lobbying efforts [F7], the Federation of Swedish Farmers (LRF) gained financial governmental support [F6] to run a pilot plant for 1G production during the first half of the 1980s [F1, F2]. In parallel, the existing production in Örnsköldsvik was supported by the government for reasons of national security of supply of chemicals (exogenous). New markets were however sought for economic reasons [F1].[24] Inspired by international developments in the field (exogenous), the focus of SSEU was to promote the use of ethanol from biomass for transport [F7]. The diversity of actors involved in SSEU [F3] resulted in the vision of 1G bio-ethanol being a bridge to the more advanced 2G bio-ethanol [F4].[25] However, voices were raised that it would be cheaper (and more efficient) to produce methanol instead of ethanol from wood [–F4].

During the second half of the 1980s, SSEU took the initiative for a field test with two ethanol buses in Örnsköldsvik [F1, F2], and then a larger fleet demonstration in Stockholm [F1, F2]. The Stockholm project was performed in co-operation with the bus manufacturer Scania and the municipal Stockholm Public Transport [F3], but was also financed by the government [F6]. Two exogenous driving forces were present in the project; air pollution, and a possible international market for Scania's ethanol buses.[26]

In general, entrepreneurs and advocates played an important role in this early stage, where a first interaction pattern already could be seen between advocacy support, resource mobilisation, entrepreneurial activities and knowledge development.

5.2. *Bus market formation and 2G R&D*

A government bill on energy policy (Swedish Ministry of Industry 1991) – the 'Three-Party Agreement' – paved the way for further expansion [F4], and two R&D programmes followed from the agreement. First, the Ethanol Research Programme [F4], which included financial support [F6] for R&D of production processes for 2G wood ethanol, mainly performed at universities [F2]. Second, the KFB (the Swedish Transport and Communications Research Board) Biofuels Programme [F4] funded projects between 1993 and 1997 [F6]. The aim was to test and demonstrate the use of ethanol as a transport fuel [F2], which implied that the ethanol bus projects were brought up to a national level [F3] and in parallel a market for ethanol city buses evolved in Swedish municipalities [F5]. The tests were performed by universities, companies and municipalities around Sweden [F1, F2], often as joint projects [F3]. The first projects of the Biofuels Programme came from a list prepared by SSEU [F7] (Arnold and Thuriaux 1997).

The development was stimulated by exogenous factors – the growing interest for environmental issues in the municipalities and concerns for air quality.[27] In Stockholm and other larger cities, 'clean vehicle organisations' were formed to manage field tests of electric cars – an exogenous factor that later turned out to be useful for the diffusion of 'clean vehicles' in general, including those running on bio-ethanol.

This episode was strongly connected to the ongoing advocacy support from LRF and SSEU. Supplemented by guidance and funding from the government, the interaction pattern related to buses in the previous episode was repeated in other municipalities, indicating cumulative causation. Fulfilment of system functions diverged gradually and even a first market started to form.

5.3. *EU entrance and tax exemptions*

Thus, when Sweden entered the EU in the middle of the 1990s, there were already some structural elements in place in the national BIS, both regarding supply and end-use technologies. The federation of Swedish Farmers (LRF) had been planning to build a 1G bio-ethanol plant to market surplus production [F1] ever since the pilot plant was closed down in the mid 1980s [–F1]. It lobbied for better economic conditions for bio-ethanol production [F7], which were finally rendered possible as the result of the Three-Party Agreement [F4]. However, the expansion of ethanol from agricultural crops was delayed owing to Sweden's entrance into the EU, an important exogenous event. The incentives for wheat ethanol production decreased because of EU subsidies for export of agricultural surplus (exogenous). In addition, tax exemptions for biofuel pilot projects needed approval from the EU.[28]

In 1997, an application [F7] for specific tax exemptions [F5] for the planned 1G bio-ethanol plant was approved by the Swedish government [F4], and the following year it was approved as a pilot plant by the EU. A contract on low blending was arranged with the fuel distributors [F5] and the plant was built by LRF [F1]. Production started in 2001.

The main connection to the previous episode came with the guidance from the Three-Party Agreement and the ongoing advocacy support by LRF. Also in this episode, the entrepreneurial function was crucial, leading to guiding decisions and even market formation. However, there is also an important hampering influence from the exogenous EU level.

5.4. *Enlarging the BIS to include cars*

For non-fleet vehicles, which are not belonging to a common depot, there was a problem that no ethanol pumps were put up if there were no ethanol cars, while no such cars could be sold if there were no pumps. Backed by SSEU,[29] the former leader of the ethanol bus project in Stockholm and a car retailer from Örnsköldsvik decided to tackle this 'chicken and egg' situation by importing flexifuel cars[30] from Ford in the US [F1]. The project was gradually expanded from 3 to 300 vehicles [F1], including an increasing involvement of the fuel distributor OK [F5].[31] The flexifuel introduction, which was partly funded by the KFB programme [F6], was successful in demonstrating that it was possible to use ethanol also for non-fleet vehicles [F2]. The cars were sold with mainly environmental arguments [F4]. In 1998, the flexifuel importers acquired the task from Clean Vehicles in Stockholm and the Swedish Energy Agency of organising a larger purchase of a new generation of flexifuel car, resulting in the introduction of Ford Focus flexifuel in 1999 [F5].[32]

This is perhaps the clearest example of how entrepreneurs set out to build up parts of the BIS that were lacking from previous episodes. Cumulative causation related to buses was now repeated for cars, with some actors (SSEU, the City of Stockholm and KFB) still involved. This was the start for a larger expansion of the use of 'clean vehicles' in Sweden.

5.5. *R&D programmes and large-scale strategies*

An ambition to create 'a green society' was pronounced by the Prime Minister in 1996 [F4]. After this, the ethanol research programme was followed up by a new one [F4], to be pursued during 1997–2004, including enlarged funding [F6] for 2G bio-ethanol R&D [F2] (Swedish Ministry of Industry and Trade 1997). In addition, a programme co-funding environmentally related investments [F6] was introduced by the Swedish Environmental Protection Agency [F4]. This contributed with funding for the later stages of the ethanol flexifuel introduction [F6].[33]

At the end of the 1990s, two exogenous factors were strengthened – climate change and rising oil prices. As a consequence, in Sweden, increasing legitimacy was seen for efficient large-scale solutions for transport fuels (e.g. gasification technology), which implied that 1G biofuels were being questioned [−F4]. Despite this controversy, there was also co-operation in lobbying for bio-ethanol in general, between advocates of 1G bio-ethanol and advocates of R&D of more efficient, 2G bio-ethanol production [F7].

In 2002, four governmental agencies (VINNOVA [Swedish Agency for Innovation Systems], Swedish Energy Agency, Swedish Environmental Protection Agency, and Swedish Road Administration) agreed on a common strategy for the introduction of biofuels, where production of 2G bio-ethanol was to be 'demonstrated by building a plant' [F4]. Blending of biofuels was generally preferred to the use of pure fuels, which should be used where they were 'economically feasible' [F4] (Eriksson et al. 2003).[34] Following this strategy, the two ethanol research programmes were followed up by a 2G pilot plant [F2], built in Örnsköldsvik. It started production in 2004, and was mainly financed by the Swedish Energy Agency [F6].[35] Though a 1G technology, wheat ethanol also survived the discussion on large scale solutions and was recommended for blending into petrol in the four agencies' strategy document [F4].

Cumulative causation was here mainly related to 2G bio-ethanol R&D. Building on previous episodes, guidance of the search was followed by resource mobilisation and knowledge development. In addition, despite the criticism, 1G bio-ethanol was still present because important system functions had been established in its favour.

5.6. *Large scale market stimulation*

Despite the criticism of 1G fuels, beginning around 2000, various economic incentives were introduced for biofuels and 'clean vehicles', such as reduced vehicle tax and lower benefit tax for privately used company cars on the national level [F5], and parking fee subsidies on the local level [F5]. General exemptions from energy and CO_2 tax for biofuels were decided upon [F5], which contributed mainly to the 5% blending of bio-ethanol in the petrol sold in Sweden (Swedish Ministry of Finance 2006). In parallel with the Swedish 1G production, the import of ethanol produced from sugar cane in Brazil was rapidly increasing (Swedish Board of Agriculture 2006).

The Swedish policies for market stimulation were further justified through the EU Biofuels Directive, and national targets were set [F4]. 'Clean vehicles' were exempted from the congestion tax in Stockholm during the trial [F5], and some national and local authorities were obliged to purchase certain shares of 'clean vehicles' [F5]. To guarantee the build-up of a refuelling infrastructure, there was a political agreement that an increasing number of filling stations should supply at least one renewable fuel, which would mainly imply bio-ethanol [F5] (Swedish Ministry of Sustainable Development 2005). Together the various measures contributed to a rapid expansion of the market for flexifuel cars and the introduction of several new models [F5].

This episode was characterised by the strengthening of the market formation function, but also–following from the new economic incentives – resource mobilisation. The emergence of these functions, however, was a result of the variety of system functions built up through cumulative causation in the previous episodes. Most system functions were strong and a take-off was awaited for domestic bio-ethanol production, though sometimes criticised mainly for economic reasons. 2G bio-ethanol was several years ahead, and would have to compete for R&D resources with

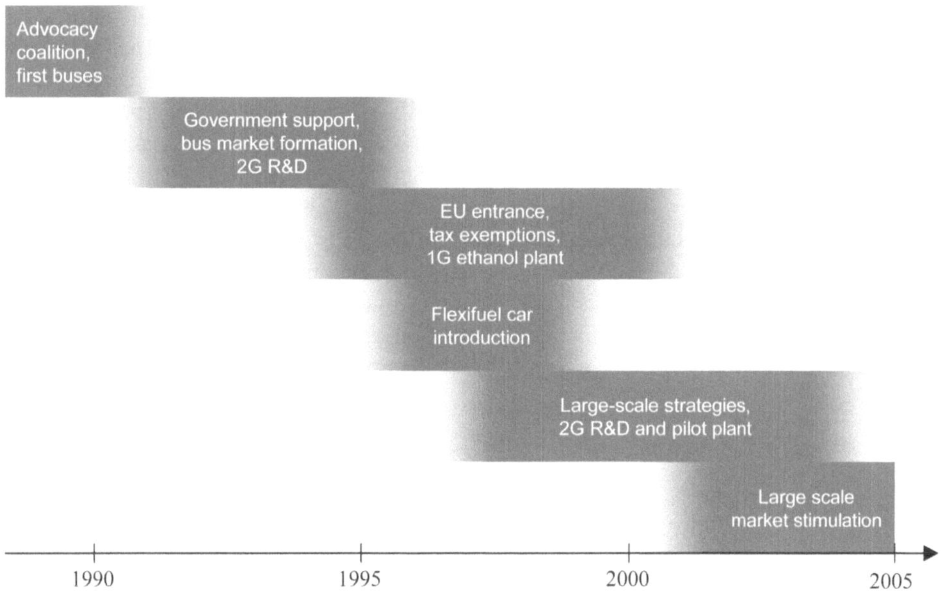

Figure 2. Six episodes of Swedish BIS development.

other 2G technologies (mainly gasification). See Figure 2 for an overview of the six episodes of Swedish BIS development.

6. Comparing the Dutch and the Swedish cases

Our cases show that Sweden and the Netherlands differ significantly regarding the development of their respective BISs. The event sequences portrayed have revealed that in the Dutch BIS, the system functions were fulfilled more or less sporadically and that cumulative causation occurred only for relatively short periods. It could be argued that since the 2003 EU directive this is changing. All system functions are present and seem to increasingly reinforce each other. In Sweden, signs of cumulative causation could be seen already at the onset of the 1990s. Since then most system functions have undergone a continuous positive development. So there are significant differences between the two countries, implying important lessons for BIS actors, such as policy makers and entrepreneurs.

6.1. *Addressing multiple system functions*

As suggested by the TIS framework, our cases confirm the importance of addressing the whole range of system functions. Policy measures are more effective when designed to target all the BIS functions, but also entrepreneurs are most successful in bringing about change when actions are directed at facilitating multiple system functions at once. This is actually what makes cumulative causation possible. Our two cases clearly differ in this respect – especially with regard to policy.

The Dutch government has shown a rather linear orientation to the promotion of biofuels. From the moment that support of 2G biofuels was deemed desirable as a policy strategy, the government has been focusing on subsidising R&D and the formation of networks, but the market was left largely unaddressed. Even the influential GAVE programme lacked attention to market formation. This was – at least partially – the reason for the discontinuation of entrepreneurial activities in the form of a commercial demonstration of 2G biofuels. The possible advantage of a broader policy orientation is shown by the variety of actions taken by the Swedish national government. Besides guidance, Swedish policy targeted knowledge formation, but also – in a more parallel approach – the early entrepreneurial experiments and the (niche) market activities around 1G biofuels. This was partly done by providing financial support but also by means of market stimulation. Influential initiatives are the bus projects supported by the KFB programme, tax exemptions for 1G bio-ethanol, and economic incentives related to vehicle purchase and ownership.

However, the orientation of Swedish policy was as much a result of, as a cause for wider BIS dynamics, and not completely intentional and strategic. Entrepreneurial activities and the formation of advocacy coalitions had a significant impact on policy making.

6.2. *The crucial role of entrepreneurs*

The TIS framework recognises the importance of entrepreneurs through the existence of an explicit function to map their activities. Though various system functions were involved in the realisation of interaction patterns in both countries, our analysis suggests that entrepreneurial activities played a crucial role in stimulating such dynamics. For most occurrences of cumulative causation at the early stages of development related to 1G, the private and municipal entrepreneurs have been prime movers, contributing to the build-up of support from advocacy coalitions, guidance of the search and other functions of the BIS.

In the Netherlands, cumulative causation has generally been short lasting, but where it occurred, it was driven by entrepreneurs. This, for instance, is true for the episode wherein Dutch boating companies started adoption experiments, raising expectations and lobbying for government support. Such activities drew in even more entrepreneurs. Also in Sweden, entrepreneurial activities around ethanol were often the start of a chain of interrelated activities. The first examples are the ethanol bus projects that started before 1990 in Örnsköldsvik and Stockholm, which opened up for further developments towards a market for ethanol city buses. A few years later, similar projects were run in the car segment, partly by the same entrepreneurs. Entrepreneurial activities (of LRF) were also a premise for the building of a 1G bio-ethanol plant.

6.3. *Reducing long-term uncertainty*

So, the entrepreneurial activities play a central role in both countries, but these alone cannot provide enough momentum. As initial policy measures were made possible by entrepreneurial activities, the continuation of entrepreneurial activities was strongly related to the presence of policy support. Our cases show that follow-ups on entrepreneurial activities were strongly related to the presence of consistent government policy. Strong programmatic policy guidance, reducing long-term uncertainty, had a sustaining effect, while the absence thereof was a recurring barrier.

Especially in the Netherlands, the absence of clear policy guidelines made promising entrepreneurial activities too risky to continue (Shell, TNO). Policy makers surely did make contributions to the fulfilment of other system functions – such as resource mobilisation (R&D subsidies) and market formation (tax exemptions), but these were not part of a general policy programme. A positive example is the programmatic support by the national GAVE programme, which strongly influenced a variety of entrepreneurs, especially around 2G technologies. The importance of guidance of the search becomes even clearer from the Swedish case; here, an important decision was taken very early, through the Three-Party Agreement in 1991. Support for ethanol use was warranted, paving the way for R&D programmes and a scaled-up use of ethanol buses. The agreement also resolved the long-standing uncertainty for 1G bio-ethanol production. Another important example is the strategy document elaborated by four governmental agencies in 2002, where a demonstration plant for 2G bio-ethanol and blending of biofuels into petrol (and diesel) was recommended. Both these Swedish examples were strongly related to functions already present in the national BIS.

In both countries the influence of policy programmes – positive or negative – has been crucial. When the government issued consistent policy guidelines, entrepreneurial activities were initiated that fostered further event sequences – and cumulative causation – in the BIS. So, the absence of a consistent policy guidance seems to result in a BIS dominated by uncertainty, with a destructive effect on further experimenting within the emerging technology field.[36]

6.4. *Policy support at different levels*

Our cases show that policy support can – and should – be provided at different levels, i.e. the regional, the national, or the EU level. In both countries the regional level has been important as a source of market-directed innovation. The earliest entrepreneurial experiments were – in several cases – related to regional companies that were strongly linked up with municipality or province. As regional initiatives were connected, a knowledge-oriented industry cluster emerged, but most importantly also advocacy coalitions that pressed national government for policy measures and legislative change. In the Dutch case this became most apparent through the coalition around 1G biofuels in the rural areas. In Sweden, the two first ethanol bus projects were started at the local level. The response from the national government level to these 'niche' dynamics has been very different in the two countries. The Swedish BIS was strengthened through a mostly facilitative orientation, fostering the regional developments and thereby contributing to the build-up of cumulative causation. In contrast, the Dutch government largely neglected support of all developments around 1G, which were mainly driven by regional interests.

An important difference also lies in the countries' relation to EU policy. In the Netherlands, national policy has been largely driven by EU incentives, though mostly it was lagging behind. As a result, the regional projects were typically stimulated by EU policy, while hampered by (the lack of) national policy. Conversely, in Sweden, the national government took the initiative of providing guidance, while the entrepreneurial activities related to 1G fuels – in the beginning of the 1990s – were held back because of the EU entrance and the change in agricultural policy. The developments were continued nonetheless, and years later – when the EU directive was launched – introduction of biofuels had already started in Sweden.

This shows the importance of support from different policy levels, but also that conflicting signals at different levels become a barrier for further developments. Where policy on the national level is lagging behind, coordinated incentives on the regional and the EU levels may stimulate national developments. Where national government shows higher ambitions, the EU policies can be troublesome.

6.5. *Hard technological choice*

From the above, we see that policy makers have been influential in framing the selection environment for biofuels technologies, both in a positive and a negative sense. Our comparison shows how the more fruitful situations were those where national policy allowed for various technologies to develop in parallel, taking into account their respective stage of development. When hard technological choices were made by policy makers, important system functions were weakened.

Government authorities in the Netherlands have been closely involved in the discussion on the technology features themselves. The 2G technologies have been presented as the group worthy of (programmatic) support, while the 1G technologies were left to the market. This way a large part of the BIS was excluded already at an early stage of development. The idea of 1G fuels being bridging technologies to 2G only comes with the implementation of the EU directive. In Sweden, there have been advocates representing either 1G or 2G alternatives, or both, whose standpoints were taken up in the policy-making process. The potential controversy between 1G and 2G was played down because of the resulting governmental support for a variety of alternatives at different stages of development.

Though this idea was not explicitly articulated by policy makers, the Swedish case suggests that cumulative causation was largely facilitated through activities related to the field of 1G bio-ethanol, where entrepreneurial activities were abundant. Also the most recent Dutch developments show that the technically less promising biofuels should not be underestimated with respect to their potential impact on cumulative causation. A direct selection on behalf of the government towards (primary) technological features yields the risk of killing off a potential for the future.[37]

7. Conclusions

Our analysis partly confirms and explains results suggested in earlier literature. However, applying the framework of Technological Innovation Systems (TISs) to the field of biofuels adds new evidence to the understanding of how emerging technologies develop under the influence of endogenous (and exogenous) forces of change. In addition, some specific lessons on biofuels can be drawn for policy makers and entrepreneurs from our comparison of two Biofuel Innovation Systems (BISs).

Taking a step back, looking at what we have learned, we can conclude that our analysis pointed out the crucial importance of dynamics when trying to understand innovation processes around technologies in a stage of emergence. We adopted the TIS framework to determine the factors that influence such processes, by analysing systematically the build-up of system functions over time. This made it possible to pinpoint how system functions interact and may lead to cumulative causation. The comparative angle allowed us to show similarities and differences from which important lessons could be derived for influencing and improving such processes in the future.

We have particularly shown how the Netherlands and Sweden differed in stimulating the build-up of a comprehensive set of system functions, necessary for the successful development of their respective Biofuels Innovation Systems (BIS). As a result of a continuity in the sequence of various entrepreneurial experiments and the emergence of cumulative causation, the Swedish BIS has reached market expansion and social implementation of biofuels, whereas the Dutch have so far established considerably less. This could only be explained to a limited extent by exogenous factors, suggesting the importance of endogenous forces – system functions – in stimulating BIS dynamics.

The TIS framework can be valuable for policymakers, but also for entrepreneurs. As long as they are active in supporting an emerging technological field, the research results may provide them with strategies to accelerate the build-up of the TIS of which they are part. (Of course, if they want to block this build-up, the results are also useful.)

For instance, both in the Netherlands and Sweden, entrepreneurial activities proved to be a crucial function as a prime mover, triggering cumulative causation. A most important difference between the two countries refers to the system function of guidance. Our study suggests that to reduce uncertainty and to provide for continuous entrepreneurial activities, policy makers, at multiple government levels, should give clear signals of desired developments and intended support programmes. Governments should not pick winners at an early stage of development, but facilitate TIS development. This can only be done by targeting multiple system functions simultaneously, implying a broad policy approach, not just aiming towards a single technology or a single part of the TIS (R&D or market). Entrepreneurs may contribute themselves by forming advocacy coalitions to provide better articulation towards governments of what they believe is desirable or feasible, but should also direct their actions at various system functions.

Some limitations should be expressed as well. Our approach uncovers particularly the dynamics that are endogenous to the technological field. Some macro-events are taken into account as exogenous forces of change. However, many differences at the macro-level between the two countries with regards to cultural, industrial and political structures and events have not been considered as possible alternative explanatory variables.

A second limitation concerns the methodology used. Since it is only in retrospect that we can acquire an insight in the consequences of particular (sequences of) events, it remains difficult to provide truly generalisable lessons. Still, this limitation is not unique to this approach but an inherent feature of social science. We see our ambition to systematise narratives, or history, using the model of a system with functions, to be in line with Freeman and Louçã's plea for reasoned history (Freeman and Louçã 1989). Hopefully, we contribute with insights and advice that are useful for policy makers and entrepreneurs in many technology fields.

Acknowledgements

We are grateful for comments given at the workshop 'Sustainable Innovation Journeys' in Utrecht, 2–3 October 2006. Special thanks to Staffan Jacobsson, Koen Frenken, Frans Berkhout and one anonymous referee for detailed comments on different versions of the paper. Volvo Research and Educational Foundations (VREF), NWO (Dutch Science Foundation), the Dutch Knowledge Network on System Innovations (KSI) and RIDE (R&D, Innovation and Dynamics of Economies) are thanked for financial support.

Notes

1. There has been an equal contribution to this paper by Hillman and Suurs.
2. For a more general account on the characteristics of the formative stage of technological systems, see Van de Ven (1993), Rip and Kemp (1998), Unruh (2002) and Jacobsson and Bergek (2004).
3. This definition is based on the one by Carlsson and Stankiewicz (1991). For an alternative definition in this special issue, see the contribution by Bergek, Jacobsson and Sandén (2008).
4. For an argument on the applicability of the innovation systems approach to the issue of energy system innovation, see also Sagar and Holdren (2002).
5. See also the more recent work by Balzat and Pyka (2006) who do a cluster analysis on the innovation system characteristics of OECD countries to classify different systems into groups.
6. They thus differ slightly from the functions used by Bergek, Jacobsson and Sandén (2008).
7. By leaving these factors outside the TIS, we run the risk of missing out dynamics 'on a higher level'. In return, this makes possible a more detailed analysis of the factors included.
8. For a historical perspective on the concept, see Skott (1994).
9. The idea of a virtuous or vicious circle is rather abstract. As we will see, the actual dynamics of a TIS, are often not exactly circular. Even under presence of positive interactions between system functions, we will find that developments

tend to be more chaotic than the ideal type of a recurring circle. Nevertheless the concept of a circle is useful as it provides a reference point to relate our findings to. We follow a rather narrow definition of what is virtuous and what is vicious dynamics; the division merely captures the dynamics of the TIS and does not (directly) relate to the desirability of the TIS or the technology it surrounds. Note that Myrdal's original concept of a vicious or virtuous circle was applied to large social systems. As we are now dealing with a far more specific problem, the development of a TIS, it could very well be that the concept does not apply in the same way. This is the reason why our research results are theoretically relatively 'loose' and mostly oriented towards close empirical observations. On the other hand the concept has been applied earlier with success; cf. Jacobsson and Bergek (2004).

10. This paper springs from two previous studies from the Netherlands and from Sweden, presented in Suurs and Hekkert (2005, 2008), and Sandén and Jonasson (2005), respectively. Both studies were based on empirical material collected from articles in periodicals and newspapers, and reports from governmental agencies, academia and consultancy firms. Literary sources were supplemented by interviews with key persons involved, and the results found were checked through discussions with industry representatives.
11. This has been empirically shown by Negro, Hekkert, and Smits 2007; Negro, Suurs and Hekkert 2008.
12. The targets were 2% for 2005 and 5.75% for 2010, of the petrol and diesel used for transport purposes (EU 2003).
13. It may seem inappropriate to consider the technology dimension an exogenous factor, but this is a matter of delineation. Adhering to the tradition of innovation system studies, we choose to focus on the societal dynamics around technology development.
14. There are also significant activities related to other biofuels. The accounts given here are for the large part based on more detailed event analyses; see Suurs and Hekkert (2005, 2008), and Sandén and Jonasson (2005), for a more elaborate account.
15. Ministry of VROM (Interview 2006), Anonymous personal communication with senior policy maker.
16. GAVE (Gasvormige en Vloeibare Energiedragers): Platform for Gaseous and Liquid Fuels; Novem (Nederlandse Onderneming voor Energie en Milieu): Dutch Agency for Energy and the Environment. The agency is mainly responsible for executing government policy on energy, innovation and sustainability.
17. GAVE (Interview 2005); personal communication with Eric van den Heuvel: Programme Manager GAVE.
18. *Ibid.*
19. *Ibid.*
20. Knowledge network, active in development of advanced bioenergy conversion technologies. The network consists of oil company Shell, ECN (Energy research Centre of the Netherlands), and numerous universities, technology institutes and energy companies.
21. TNO is the Dutch Organization for Applied Scientific Research. The network consists of TNO, a number of companies active in chemical engineering and energy distribution.
22. Ministry of Economic Affairs (Interview 2006); personal communication with a Senior Policy Maker; GAVE, see note 18.
23. The early adoption experiments are all linked to foreign supply.
24. Foundation for Swedish Ethanol Development, previously SSEU (Interview 2005); personal communication with S. Flodin.
25. *Ibid.*
26. Protima AB (Interview 2005); personal communication with C. Rydén.
27. The interest was partly associated with the Agenda 21 framework after the Rio World Summit in 1992.
28. The Federation of Swedish Farmers (LRF) and Lantmännen (Interview 2005); personal communication with E. Herland.
29. About this time, SSEU was renamed BAFF (BioAlcohol Fuel Foundation).
30. Flexifuel vehicles can run on any mix of petrol and E85, which consists of 85% ethanol and 15% petrol.
31. Protima AB, see note 27.
32. *Ibid.*
33. *Ibid.*
34. The strategy was mainly in support of gasification technology producing other fuels than ethanol.
35. In parallel, pilot plants for gasification technology (i.e. not related to ethanol) – were prepared.
36. Note that the guidance function need not necessarily be fulfilled by policy makers. In fact, entrepreneurs themselves can play an important role, for instance, by collaborating through the formation of advocacy coalitions. However, for the long-term perspective it turns out that government leadership is a requirement.
37. The development of various technologies can only be stimulated through the use of several different policy instruments in parallel. See Sandén and Azar (2005) on the need for technology differentiated support. Of course, there is need for a balanced view, as too much focus on 1G technologies may lead to a dead end. See Hillman and Sandén (in press).

References

Algemeen Dagblad (*AD*). 1992. Milieu gebaat bij bio-ethanol als motorbrandstof. *AD*, 7 May 1992.

Arnold, E., and B. Thuriaux. 1997. The biofuels programme at KFB, 1991–97: an evaluation. Kommunikationsforskningsberedningen (KFB), Stockholm.

Balzat, M., and A. Pyka. 2006. Mapping national innovation systems in the OECD area. *International Journal of Technology and Globalisation* 2, no. 1/2: 158–76.

Bergek, A. 2002. Shaping and exploiting technological opportunities: the case of renewable energy technology in Sweden. PhD thesis, Department of Industrial Dynamics, Chalmers University of Technology, Göteborg.

Bergek, A., S. Jacobsson, and B.A. Sandén. 2008. 'Legitimation' and 'development of external economies': two key processes in the formation phase of technological innovation systems. *Technology Analysis & Strategic Management*, this issue.

Bergek, A., S. Jacobsson, B. Carlsson, S. Lindmark, and A. Rickne. 2008. Analyzing the functional dynamics of technological innovation systems: a scheme of analysis. *Research Policy* 37, no. 3: 407–29.

Bijker, W.E., T.P. Hughes, and T. Pinch, eds. 1987. *The social construction of technological systems*, Cambridge, MA: MIT Press.

Bizz. 2002. Olie in de autotank. *Bizz*, 18 October 2002.

———. 2002b. De Oliemolen (2). *Bizz*, 15 November 2002.

Boerrigter, H., H. Den Uil, and H.-P. Calis. 2002. Green diesel from biomass via Fischer-Tropsch synthesis: new insights in gas cleaning and process design. Paper presented at the Pyrolysis and Gasification of Biomass and Waste Expert Meeting, 30 September–1 October 2002, Strasbourg, France.

Breschi, S., F. Malerba, and L. Orsenigo. 2000. Technological regimes and schumpeterian patterns of innovation. *Economic Journal* 110, no. 463: 388–410.

Carlsson, B., and R. Stankiewicz. 1991. On the nature, function and composition of technological systems. *Journal of Evolutionary Economics* 1, no. 2: 93–118.

Carlsson, B., S. Jacobsson, M. Holmén, and A. Rickne. 2002. Innovation systems: analytical and methodological issues. *Research Policy* 31, no. 2: 233–45.

Dagblad van het Noorden. 2004. Accijnsvrij, *DN*, 11 August 2004.

———. 2005. Lesauto op biodiesel: Hij stinkt niet, hij ruikt anders. *DN*, 17 March 2005.

Edquist, C. 2005. Systems of innovation: perspectives and challenges. In *The Oxford handbook of innovation*, eds. J. Fagerberg, D.C. Mowery and R.R. Nelson, 181–208. Oxford: Oxford University Press.

EIA. 2003. *International energy outlook 2003*. Energy Information Administration, US Department of Energy, Washington, 2003.

Eriksson, P., T. Korsfeldt, L.-E. Liljelund, and I. Skogö. 2003. Introduction of biofuels on the market – the public administration reference group recommendations. Swedish Road Administration, Swedish Energy Agency, VINNOVA (Swedish Agency for Innovation Systems), Stockholm.

EU. 1997. Energy for the future: renewable sources of energy – White Paper for a community strategy and action plan, COM(97)599 final. European Commission.

———. 2003. Directive 2003/30/EC of the European Parliament and of the Council – of 8 May 2003 – on the promotion of the use of biofuels or other renewable fuels for transport. *Official Journal of the European Union*, L 123: 42–6.

Freeman, C. 1995. The 'National System of Innovation' in historical perspective. *Cambridge Journal of Economics* 19, no. 1: 5–24.

Freeman, C., and F. Louçã. 2001. *As time goes by: from the industrial revolutions to the information revolution*. Oxford: Oxford University Press.

GAVE. 1999. Een energiek klimaat voor neutrale dragers. Eindadvies van de inventarisatie van het GAVE-programma. SenterNovem, Utrecht, The Netherlands.

———. 2002. Overzicht projectenprogramma GAVE-2001. *Novem GAVe-mail*, no. 1.

———. 2003. Nieuwe rapporten op de GAVE website. *Novem GAVe-mail*, no. 11.

Geels, F.W. 2005. Processes and patterns in transitions and system innovations: refining the co-evolutionary multi-level perspective. *Technological Forecasting & Social Change* 72: 681–96.

Hamelinck, C.N., and A.P.C. Faaij. 2006. Outlook for advanced biofuels. *Energy Policy* 34, no. 17: 3268–83.

Hekkert, M.P., R.A.A. Suurs, S.O. Negro, S. Kuhlmann, and R.E.H.M. Smits. 2007. Functions of innovation systems: a new approach for analysing technological change. *Technological Forecasting & Social Change* 74: 413–32.

Het Financieele Dagblad. 1995. Biodiesel tijdelijk vrij van accijns. *FD*, 8 March 1995.

Hillman, K.M., and B.A. Sandén. 2008, in press. Exploring technology paths: the development of alternative transport fuels in Sweden 2007–2020. *Technological Forecasting & Social Change*.

IEA. 2004. *Biofuels for transport – an international perspective*. Paris: International Energy Agency & Organisation for Economic Co-operation and Development.

———. 2006. *World energy outlook 2006*. Paris: International Energy Agency & Organisation for Economic Co-operation and Development.

Jacobsson, S., and A. Bergek. 2004. Transforming the energy sector: the evolution of technological systems in renewable energy technology. *Industrial and Corporate Change* 13, no. 5: 815–49.

Jacobsson, S., and A. Johnson. 2000. The diffusion of renewable energy technology: an analytical framework and key issues for research. *Energy Policy* 28, no. 9: 625–40.

KEMA Consultancy. 2000. Optimale inzet van biomassa voor energieopwekking. GAVE Report no. 2, Arnhem, The Netherlands.

Landbouw-Economisch Instituut. 2005. *Beschikbaarheid koolzaad voor biodiesel*. Wageningen: LEI.

Leeuwarder Courant. 2004. Koolzaadauto's ruiken naar frituurvet. *LC*, 19 October 2004.

Myrdal, G. 1957. *Economic theory and under-developed regions*. London: Ducksworth.

Negro, S.O., M.P. Hekkert, and R.E. Smits. 2007. Explaining the failure of the Dutch innovation system for biomass digestion: a functional analysis. *Energy Policy* 35, no. 2: 925–38.

Negro, S.O., R.A.A. Suurs, and M.P. Hekkert. 2008. The bumpy road of biomass gasification in the Netherlands: explaining the rise and fall of an emerging innovation system. *Technological Forecasting & Social Change* 75, no. 1: 57–77.

NRC Handelsblad. 1992. Vierde gewas moet eenzijdige akkerbouw redden. *NRC*, 8 August 1992.

Poole, M.S., A.H. Van de Ven, K. Dooley, and M.E. Holmes. 2000. *Organizational change and innovation processes: theory and methods for research*. Oxford: Oxford University Press.

Provinciale Zeeuwse Courant. 2004. Biodieselfabriek in Zeeuws-Vlaanderen kan in 2005 draaien. *PZC*, 12 August 2004.

Rip, A., and R. Kemp. 1998. Technological change. In *Human choice and climate change*, vol. 2, 327–399. Columbus, OH: Battelle Press.

Rotterdams Dagblad. 2004. Regio positief over schonere voertuigen: alle gemeentelijke wagenparken over op bio-ethanol. *RD*, 20 December 2004.

Sagar, D., and J.P. Holdren. 2002. Assessing the global energy innovation system: some key issues. *Energy Policy* 30, no. 6: 465–9.

Sandén, B.A., and C. Azar. 2005. Near-term technology policies for long-term climate targets – economy wide versus technology specific approaches. *Energy Policy* 33: 1557–76.

Sandén, B.A., and K.M. Jonasson. 2005. Variety creation, growth and selection dynamics in the early phases of a technological transition: the development of alternative transport fuels in Sweden 1974–2004. ESA-report 2005: 13, Environmental Systems Analysis, Chalmers University of Technology, Göteborg, Sweden.

Skott, P. 1994. Cumulative causation. In *The Elgar companion to institutional and evolutionary economics*, eds. G.M. Hodgson, W.J. Samuels and M.R. Tool. Cheltenham: Edward Elgar.

Stromen. 2003. Alleen door verplichtstelling kunnen biobrandstoffen doorbreken in Nederland. *Stromen*, 28 February 2003.

———. 2004. Nog geen haast met stimulering van biobrandstoffen. *Stromen*, 5 November 2004.

Suurs, R.A.A. and M.P. Hekkert. 2005. Naar een methode voor het evalueren van transitietrajecten; functies van innovatiesystemen toegepast op biobrandstoffen. Report Department of Innovation Studies, Utrecht University.

———. 2008. Cumulative causation in the formation of technological innovation system: the case of biofuels in the Netherlands. Innovation Studies Utrecht (ISU). Working Paper, Series no. 08-04.

Swedish Board of Agriculture. 2006. Marknadsöversikt: Etanol, en jordbruks-och industriprodukt. Jönköping, Sweden.

Swedish Ministry of Finance. 2006. Skattebefrielse för biodrivmedel. Swedish Government, Stockholm. http://www.regeringen.se

Swedish Ministry of Industry. 1991. Government Bills: Prop. 1990/91:88. Swedish Government, Stockholm.

Swedish Ministry of Industry and Trade. 1997. Government Bills: Prop. 1996/97:84 – En uthållig energiförsörjning. Swedish Government, Stockholm.

Swedish Ministry of Sustainable Development. 2005. Government Bills 2005: Prop. 2005/06:16 – Skyldighet att tillhandahålla förnybara drivmedel. Swedish Government, Stockholm.

Unruh, G.C. 2000. Understanding carbon lock-in. *Energy Policy* 28, no. 12: 817–30.

———. 2002. Escaping carbon lock-in. *Energy Policy* 30, no. 4: 317–25.

Van de Ven, A.H. 1993. The development of an infrastructure for entrepreneurship. *Journal of Business Venturing* 8: 211–30.

Discourse and innovation journeys: the case of low energy housing in the UK

Heather Lovell

The paper examines the role of discourse in innovation journeys using the example of low energy housing in the UK. Discourse is shown to be influential within innovation journeys in two main ways: first, discourse unites the often disparate organisations involved in innovation, and thereby gives structure and direction to the innovation journey; second, discourse has the power to retrospectively 'reframe' the course of an innovation journey, leaving out inconsistent parts and ignoring twists and turns, so past innovation journeys are in effect reconstructed in the present. This discursive remapping of innovation journeys has implications for current and future pathways of innovation. Further, there is a two-way relationship between discourse and technology development. It is discussed how pioneering low energy housing developments have themselves become a significant part of low carbon housing discourse, acting as powerful 'storylines'.

Introduction

The main objective of the paper is to assess the key processes, mechanisms and patterns in an innovation journey from a discourse analysis perspective, using the case of low energy housing in the UK. Discourse analysis incorporates a large literature (see Dryzek 1997; Sharp and Richardson 2001 for an overview), and for this reason the paper focuses on two key discursive processes within the UK low energy housing innovation journey: first, the emergence of a network of actors united by language about climate change, a low carbon housing discourse coalition (Hajer 1995); and second, the framing and reframing of innovations as solutions to different policy problems (Laws and Rein 2003; Rein and Schon 1993). It is shown how discourse is particularly important during the developmental stage of an innovation journey because once prototypes have been built, and innovations are in the public arena and have a visible material presence, it becomes possible for organisations not involved in the initiation stage of an innovation journey to try to claim ownership of the innovation through discursively reframing it to meet their objectives.

The role of discourse in innovation journeys has not been examined in detail, which is surprising because the notion of an innovation journey concerns ideas, people and relationships, and thereby has significant overlap with discourse analysis, which concentrates on similar themes (Dryzek 1997; Flyvbjerg 1998; Hastings 1999; Rydin 1999; Sharp and Richardson 2001). An innovation journey is defined by Van de Ven et al. (1999, 6) as 'new *ideas* that are developed and implemented to achieve desired *outcomes* by *people* who engage in *transactions* (relationships) with others in changing institutional and organizational *contexts*' (emphasis in original). It is argued here that the politics and power struggles highlighted through a focus on discourse are integral to the innovation process, and deserve more attention. Discourse both constrains and enables innovations, in particular through its role in simplifying the messiness of innovation journeys.

An objective of the paper is therefore to broaden Van de Ven's original approach to innovation journeys, which focused on corporations and discrete products (Van de Ven et al. 1999). The empirical focus of the paper is on the changing policy and socio-technical context within which a diverse group of technical and administrative low energy housing innovations developed in the UK from the 1970s to 2006–2007, including photovoltaic panels, solar design, insulation, wind turbines, micro-CHP plants, self-build housing, and new models of planning and designing housing. It is not the author's intention to examine in detail the journey of each specific innovation. Rather, analysis concentrates on the influence of discourse on low energy housing innovations as a whole, as well as the effect of the low energy housing innovations themselves on discourse and policy. The paper explores how low energy housing innovations were initially developed and trialled in the UK within the sustainable housing movement during the 1970s. Sustainable housing – originally designed to meet a number of broadly defined environmental and social objectives – became reframed more narrowly as 'low carbon housing' by a loosely associated group of policy actors – a discourse coalition – in the late 1990s as climate change rose up the UK policy agenda. The process of low carbon reframing has significantly remapped the historical innovation journey, and this in turn has had a bearing on the contemporary pathway of low energy housing technologies.

The analysis is based on a combination of interview and documentary evidence collated through a UK Economic and Social Research Council (ESRC) funded doctorate (2001–2005), and Post-doctoral Fellowship (2005–2007). Approximately 50 in-depth semi-structured interviews were conducted with a range of people involved in sustainable housing, climate change policy and low carbon technology innovation in the UK. Organisations interviewed included local and national government, sustainable housing groups, consultancies, Registered Social Landlords, non-governmental organisations, regional government agencies, architects, renewable energy firms, and private sector house builders. The interviews were transcribed and coded in order to identify key discourses. In addition, documentary evidence was compiled and analysed from a range of sources including government policy documents, housing and energy industry trade magazines, and the national press.

The paper is structured as follows. First, the role of discourse in innovation journeys is assessed, and theories about discourse coalitions and discursive framing are introduced. Second, the UK low energy housing innovation journey is explored using a discourse analysis approach, from its early origins in the sustainable housing movement of the 1970s, to the later dominance of a low carbon housing discourse coalition centred on the problem of climate change. It is discussed how the low carbon discourse coalition has used ecologically modern discourse to promote the benefits of low energy housing. At its simplest, ecological modernisation is the idea that there is no conflict between economic growth and protecting the environment because 'smart' green technologies can fulfil both objectives (Mol and Spaargaren 2000; Murphy 2000; Young 2000).

The metaphors or 'storylines' (after Hajer 1995) used by the low carbon discourse coalition to help explain and define their work hence have a strong technology focus. In conclusion, the author reflects on the strengths and limitations of discourse analysis to the innovation journey concept.

The relationship between discourse and innovation journeys

As mentioned, the role of discourse has not been explored in detail in the context of innovation journeys (see Van de Ven et al. 1999), so the key objective here is to highlight in what ways it might be relevant. It is argued that discourse is important in innovation journeys for two main reasons. First, it unites disparate actors involved in innovation, and thereby gives coherence to otherwise loosely co-ordinated, messy pathways of change: discourse gives structure to the innovation journey. Second, and related, discourse has the power to retrospectively change the course of innovation journeys; to 'reframe' the journey, simplifying it, reshaping or forgetting its origins, and ignoring twists and turns. As is shown to be the case with low energy housing, this retrospective remapping of past innovation journeys can affect the contemporary trajectory of innovations. These ideas are discussed with reference to two political science theories about discourse: discourse coalitions, and discursive framing.

There is a long tradition in political science of theorising policy change as driven by networks of actors – termed policy networks – mixed groups of people involved in the policy process from government, corporations, the media and non-governmental organisations (Marsh and Rhodes 1992; Sabatier 1999). Policy networks are an important influence on the direction and pace of innovation journeys because it is through these networks that decisions are made on issues such as the regulatory environment, grant funding and consumer markets for new innovations; factors all identified as critical to innovation journeys (Van de Ven et al. 1999). The policy network literature is large, and it is not my intention to discuss it in full here. Rather, I wish to concentrate on one particular type of policy network – the discourse coalition – and illustrate how discourse coalitions give structure to the messiness of innovation journeys. Members of discourse coalitions do not necessarily share the same values, rather they are driven by a variety of different aims and beliefs, and united instead by their shared use of language (Hajer 1995). The 'glue' that binds these actors is discourse – the way they talk about an issue and the metaphors or 'storylines' that they use. Storylines are defined as '*the essential discursive cement* that creates communicative networks among actors with different or at best overlapping perceptions and understandings' (Hajer 1995, 63, emphasis added). Discourse coalitions are loose, fluid networks and it is storylines that facilitate understanding between network members who do not otherwise have much in common.

Discourse coalition theory was originally developed in relation to public policy debates, rather than private sector innovation (Flyvbjerg 1998; Sharp and Richardson 2001). There are hence some limitations in translating the concept to innovation journeys, but the concept of a discourse coalition was always designed to embrace a wide range of actors – including the private sector and non-governmental actors – not just government (Bulkeley 2000; Hajer 1995; Owens and Cowell 2002). It is a broader, more inclusive concept than an industrial lobby, for example, and therefore more appropriate to the empirical case of low energy housing. Further, in this paper it is shown how the innovation journey of low energy housing has been closely connected with government and public policy as climate change has risen up the UK policy agenda, so the public policy origins of discourse coalition theory has relevance.

Discursive framing concerns how new ideas are presented, and in particular how problems and solutions are simplified in order to make them understandable and amenable to debate, it is defined as: '... a way of selecting, organising, interpreting and *making sense of a complex reality* to provide guideposts for knowing, analysing, persuading and acting' (Rein and Schon 1993, 146, emphasis added), or 'a particular way of representing knowledge, and ... the reliance on *schemas that bound and order a chaotic situation*, facilitate interpretation and provide a guide for doing and acting' (Laws and Rein 2003, 173, emphasis added). Discursive frames are hence a critical influence on innovation journeys because of their role in neatening and simplifying individual innovation pathways, retrospectively making it appear that there was only ever one predefined journey, when in reality there were multiple often overlapping innovation trajectories (Van de Ven et al. 1999). Framing sets the boundaries around an issue, or technology, and allows ownership of it by certain actors. It is thereby closely connected to the concept of a discourse coalition. There has been extensive academic debate about the relationship between discourse coalitions, frames and storylines, which is not appropriate to enter in detail here (see Rein and Schon 1993). For this analysis I assume that framing is a technique used by discourse coalitions to help define their boundaries of activity and lend coherence to their discourse. For example, it is discussed below how the low carbon housing discourse coalition draws on an ecologically modern discursive frame.

Framing is usually discussed in terms of framing an emerging situation or issue as a problem, as follows:

> A frame is a perspective from which an amorphous, ill-defined, *problematic situation* can be made sense of and acted on. (Rein and Schon 1993, 146, emphasis added)

However, framing of *solutions* can also occur, and this is what happened with UK sustainable housing: existing technologies were retrospectively framed as a solution to new policy problems that emerged during the course of the innovation journey, most notably climate change. This type of reframing occurs most readily when an innovation is disseminated within the public domain, i.e. once it has a material presence – and thus is recognised and adopted by different organisations.

In sum, considering ideas about discourse coalitions and discursive framing and reframing helps us to better recognise and acknowledge the non-linearity or 'messiness' of innovation journeys. Discourse acts to filter out and simplify seemingly irrelevant issues, technologies and ideas in order to create an effective, understandable message, thereby simplifying innovation journeys retrospectively. Analysts of innovation journeys must therefore play close attention to how discourse operates, especially to reinstating some of the complex dynamics of innovation journeys; a task turned to below.

The UK low energy housing innovation journey

This section is structured as follows. First, the origins of the low energy housing innovation journey in the 1970s are discussed, and a period of policy flux during the late 1990s is explored when policy makers became interested in sustainable housing as a solution to a number of persistent policy problems. Second, the process of 'low carbon' reframing of existing low energy housing innovations is assessed: the low carbon discourse coalition is described, and it is shown how the coalition has simplified and remapped the past innovation journey of low energy housing through use of ecologically modern discourse. 'Storylines' used by the discourse coalition about innovative housing developments – in particular two UK developments called BedZed and Hockerton – are critically assessed. These technology-based storylines have helped inspire further innovation and action, thereby speeding up and facilitating the innovation journey. However, there have been some problems arising from the low carbon discourse coalition's simplification and remapping of the low energy housing innovation journey, especially the way in which it has largely ignored the social context in which the low energy housing innovations were originally developed.

The 1970s: the early origins of UK sustainable housing

A general pattern identified in the innovation journey of low energy housing is a shift from early stage innovation in the 1970s being dominated by a single close-knit network of people with shared environmental and social values (an 'advocacy coalition', see Sabatier and Jenkins Smith 1993), to looser multiple networks of actors united by shared language – discourse coalitions – emerging during the 1990s. The case of low energy housing thus has parallels with other research regarding the professionalisation of environmental groups over time as new types of mainstream actor become involved once sustainable innovations are established (Lounsbury et al. 2003; Smith 2007). The sustainable housing movement emerged in the early 1970s in the UK (Barton 2000; Chappells and Shove 2000; Smith et al. 1998), concurrent with an increased public awareness of environmental issues, and an upsurge in radical deep green environmentalism (Dryzek 1997; Porter and Brown 1996). Examples of sustainable housing developments from this period include the Centre for Alternative Technology (CAT) in Wales and the Findhorn Ecovillage in Scotland (CAT 1995; Findhorn Ecovillage 2003). Housing developments were typically broadly designed as a solution to a number of environmental and social problems, ranging from water pollution to fuel poverty (CAT 1995; Wood 1990). Innovation was not located within private companies but rather dispersed across a range of largely informal organisations and individuals, including community groups, non-governmental organisations, green architects and social entrepreneurs (see Chappells and Shove 2000; Lovell 2004; Smith 2004). Table 1 gives some examples of key organisations and individuals involved in sustainable housing during this early stage of innovation.

The sustainable housing movement believed radical societal changes were necessary in order to achieve sustainable development, such as governance via small-scale self-sufficient communities (Dobson 2000). Crucially, those involved perceived themselves as a social movement united by deep green environmental and social values (Lovell 2004; Smith 2003, 2004). Somewhat paradoxically for this analysis, therefore, the 'deep green' beliefs that united these early innovators centred on a rejection of a technical fix to environmental problems, instead advocating a fundamental shift in attitudes and consciousness (Dobson 2000). Prototype household-level renewable energy technologies such as micro hydro, wind turbines, wall and roof insulation, etc., were adopted primarily because they enabled this network of early innovators to fulfil their desire for autonomy from modern society (Wood 1990). However, as explained in an assessment of the history of the CAT development in Wales, there were a mix of views in the early sustainable housing movement about their role as innovators – the situation was not clear-cut:

> There was always a tension [at CAT] between those who wanted to raise a drawbridge to the outside world and those who believed that what they were doing was primarily to serve others. (Richard St George, co-founder of CAT, quoted in CAT 1995, 9)

Nevertheless, despite this internal tension, these actors played an important role in the early development and testing of new low energy housing technologies (see BRECSU 1996; CAT 1995; Wood 1990).

Two issues about the relationship between discourse and innovation journeys are worth emphasising here. First, that the original context in which low energy housing innovations were developed

Table 1. Principle organisations involved in the UK sustainable housing movement in the 1970s.

Name of actor	Brief description
Centre for Alternative Technology (CAT), Machynlleth, Wales 1973+	A community development in an old quarry site in rural Wales, established in 1973. It is a self built autonomous development (energy and water self sufficient). It also operates as a sustainable housing education and resource centre, and runs residential courses (see CAT 1995). CAT publishes a quarterly sustainability magazine called *Clean Slate*
Findhorn Ecovillage, Scotland 1962+	A self built community in rural Scotland. It was established in 1962, and building on site started in the early 1970s. As with CAT, there is an education centre and residential courses (see Findhorn Ecovillage 2003)
Communes Network and Diggers and Dreamers 1968+	The Communes Network started as the Communes Movement in 1968, founded by the Selene Community in Wales. Among other activities it produced a bi-monthly magazine *The Communes Journal*, which had a print run of 3000 copies. In 1975 it became the *Communes Network*, a more loosely connected organisation, which still operates informally today. Some members of the Communes Network have formed 'Diggers and Dreamers' – a self build community organisation which aims to help self builders to network, and to access information on self build housing (Dawling 1992)
Undercurrents magazine 1972–early 1980s	*Undercurrents* was a radical environmental magazine published in the 1970s and early 1980s (commenced in 1972), and was regarded as the alternative movement's journal during this period. Its subtitle was 'the magazine of radical science and people's technology'. It focused in particular on sustainable housing communities active during the 1970s
Robert and Brenda Vale 1970s+	The Vales' both studied architecture at Cambridge University in the early 1970s, and in 1975 published 'The Autonomous House', an influential text for the sustainable housing movement. They subsequently built their own autonomous house in the village of Southwell in the East Midlands, UK, and were then the architects for the nearby Hockerton Housing Development. They remain active in the field of sustainable and low energy housing

and experimented with is very different to how they are used and promoted currently in the early twenty-first century. Those using low energy innovations at the household and community scale in the 1970s were part of a radical 'alternative' social movement, united by strong values and a desire to be self-sufficient (Chappells and Shove 2000; Dobson 2000; Lovell 2004; Smith 2004). The example of low energy housing illustrates the messiness and unpredictability of an innovation journey: those organisations and individuals involved at the early stages have not been a long-term feature of the innovation journey. Second, and related, the policy context in which these low energy housing innovations developed was very different to the contemporary situation. Energy efficiency and renewable energy were issues on the government's policy agenda in the 1970s, but not because of climate change – which had yet to become widely recognised in policy circles – but largely because of fuel shortages and the OPEC oil crisis (see Toke 2000). So a low carbon – or even a low energy – innovation journey was not planned from the outset, those involved at the time could not envisage where their activities might lead: the final destination of the innovation journey was unknown. Innovations centred initially on a wide range of environmental and social issues, and it was only subsequently that low energy innovations were prioritised by other more mainstream organisations.

The late 1990s: the reframing of sustainable housing

During the late 1990s and turn of the century there was a growing interest in sustainable housing from mainstream organisations, in particular the UK government. There emerged during this period multiple discourse coalitions who framed sustainable housing as a solution to a number of policy problems, ranging from modernisation of the construction industry, to meeting the demand for new housing (see Table 2).

The empirical material in Table 2 indicates how there were multiple possible innovation pathways at this point, centred on different innovations used in sustainable housing developments, ranging from factory-based housing technologies (a solution within the 'innovation in the construction industry' discourse, see The Housing Forum 2001), to car sharing initiatives and housing without car parking (innovations adopted by the 'traffic congestion' discourse coalition, see DEFRA 2000). In other words, there were a number of different directions the innovation journey could take, with a high degree of uncertainty and messiness. It is not my intention to imply here that none of these pathways were subsequently followed: in the early twenty-first century sustainable housing continues to be framed as a solution to a number of policy problems in the UK, in particular construction industry modernisation and meeting the demand for new housing (see Barker 2004; ODPM 2003b). The situation is thus in keeping with the notion of a non-linear, contested innovation journey with multiple pathways. However, as climate change rose up the UK political agenda it was the low energy innovations that became the dominant focus of attention. Other pathways have subsequently been selectively 'edited out' by low carbon discourse coalition, as explored below, because a clear uniting message was required to unite the low carbon discourse coalition, and any 'messiness' in the story of low energy housing, such as experimentation with other social and environmental innovations, was unhelpful in this regard.

A critical influence on the low energy housing innovation journey during this period was that a reasonably large amount of sustainable housing already existed in the UK: the innovations had a material presence. Once an innovation has undergone the initiation stage of an innovation journey and is within the public domain it is more readily able to be discursively reframed in this way. Thus the material presence of the UK sustainable housing developments rendered them liable to 'capture' and discursive reframing (Lovell 2004). The developmental stage of an innovation journey – when working prototypes are within the public domain – is hence a critical period from a discourse analysis perspective because multiple future pathways for the innovations become possible as ownership of the technology is wrested from those actors originally involved in its initiation. The prior material existence of innovations makes it appear that rapid progress has been made on an issue, and thus is particularly appealing in policy areas where there is a gap between government targets and actual policy achievements. The material presence of innovations also helps demonstrate that the new technologies work, and thus reduces risk for more mainstream institutions who wish to use them. Furthermore, it helps to ground rhetoric or discourse about a particular problem and thus lends the discourse coalition credibility.

Table 2. The framing of sustainable housing as a solution to different UK policy problems (adapted from Lovell (2004)).

Policy problem sustainable housing was framed as a solution to	Organisations involved in framing	Examples of policies, housing developments and technologies
Meeting the demand for new housing	UK government; private sector house builders; local authority planners; the Town & Country Planning Association; WWF.	*Policies:* The 2003 Sustainable Communities Plan; The Millennium Communities Programme; WWF One Million Sustainable Homes campaign
		Housing Developments: 'Zed squared' zero energy, zero waste development in the Thames Gateway; the West Stevenage development; Ashton Green, Leicester
		Technologies: prefabricated housing components, high density design, community-level provision of utility systems (water, energy, waste)
Lack of innovation in the construction industry	UK government; government-funded housing innovation organisations (Rethinking Construction and the Housing Forum).	*Policies:* Speech by Construction Minister Brian Wilson April 2003 – green housing and housing sector modernisation (DTI 2003); The Housing Forum 2001 Off Site Manufacture report – 'Homing in on Excellence'
		Housing Developments: Greenwich Millennium Village, London; INTEGER housing projects at Maidenhead & Sandwell
		Technologies: 'smart' metering, prefabricated housing components, household renewable energy technologies, e.g. PV roof tiles
Poor quality of existing housing stock	UK government; social housing organisations; local authorities.	*Policies:* The Sustainable Communities 'Pathfinder' Regeneration Areas; The 2003 Decent Homes policy; Urban Regeneration Companies
		Housing Developments: inner city Tower Block refurbishment – e.g. Glastonbury House, Pimlico, London
		Technologies: low 'e' insulated windows, solar hot water systems, wall and roof insulation

Fuel poverty	Local authorities; government-funded energy efficiency advice centres; energy utilities; campaign organisations.	*Policies:* The 2001 Fuel Poverty strategy; The 2003 Energy White Paper; The 2000 Home Energy Conservation Act *Housing Developments:* Boughton Energy Village, Newark and Sherwood District Council, East Midlands; Ravenscliffe, Bradford, North British Housing Association, 64 low energy timber frame houses *Technologies:* low 'e' insulated windows, solar hot water systems, wall and roof insulation, community and micro-CHP
Traffic congestion	UK government; local authority planners; campaign organisations, e.g. Transport 2000.	*Policies:* The UK Transport 10 Year Plan (2000) *Housing Developments:* Slateford Green, Edinburgh; BedZed, south London *Technologies:* community electric car share schemes, high-density design, no car parking spaces for residents
Meeting renewable energy generation targets	Energy utilities; UK Government; renewable energy companies.	*Policies:* The 'Clear Skies' community and household grant programme; the social housing Generating Solar Homes Programme *Housing Developments:* North Nines, Edmonton, London; Sherwood Energy Village, East Midlands *Technologies:* solar hot water systems, PV panels, small wind turbines, community and micro-CHP
Climate Change mitigation ('low carbon')	Energy utilities; UK government; local authorities; Private sector house builders; social housing organisations; green architects; WWF.	*Policies:* The 2003 Energy White Paper; 'Energy - The Changing Climate'. The UK Royal Commission on Environmental Pollution's 22nd Report *Housing Developments:* The Vales' Autonomous House; Hockerton Housing Development, Notts.; BedZed, south London; Greenwich Millennium Village *Technologies:* low 'e' insulated windows, solar hot water systems, wall and roof insulation, PV panels, small wind turbines, community and micro-CHP

The reframing of sustainable housing as a solution to climate change

The remainder of the paper concentrates on how one particular discourse coalition active during the late 1990s – the low carbon discourse coalition – has come to dominate policy debate since the turn of the century (Turnpenny et al. 2005; Vidal 2006). Low energy housing innovations such as solar hot water heating systems, self build community schemes and passive solar design – initially developed and tested by the sustainable housing movement prior to the problem of climate change rising up the UK's policy agenda – have become reframed as 'low carbon'.

The current UK climate change policy framework is briefly introduced, followed by a discussion of how the discourse coalition has conducted low carbon framing using ecologically modern discourse. Technologies are an integral part of this discourse, and it is shown how certain sustainable housing developments have themselves become important low carbon 'storylines' (after Hajer 1995). The implications of the emergence of a UK low carbon discourse coalition for the innovation journey of low energy housing are considered. It is shown how the framing of sustainable housing as low carbon has speeded up the innovation journey of low energy household technologies. However, the narrowing of debate to focus on the energy aspects of sustainable housing has been contested, and some of the inconsistencies and difficulties in maintaining a coherent low carbon discourse are discussed.

Climate change policy in the UK: an introduction

Climate change has become an important policy problem within the UK and other industrialised countries since it first captured widespread public attention at the Rio 1992 Earth Summit (Bulkeley and Betsill 2003; Newell 2000). Since the turn of the century in particular climate change has risen dramatically up the UK's policy agenda (House of Commons 2006; Stern 2006; Vidal 2006). Through the international Kyoto Protocol the UK Government is committed to reducing greenhouse gas emissions by 12.5% by the year 2012, and it also has a long-term legally binding goal to lower carbon emissions by 60% by 2050 through the 2006 Climate Change and Sustainable Energy Act (HM Government 2006). It has been suggested that discourse structuration (after Hajer 1995, 60–61) has occurred with the issue of climate change in the UK, defined as when a certain way of discussing an issue dominates policy discourse, such that those not using particular phrases or storylines risk losing credibility (see Turnpenny et al. 2005).

Although the 2003 UK Energy White Paper reframed energy policy around the issue of climate change, in particular concentrating on the action required in the residential sector (DTI 2003), the government has since faced mounting criticism for its failure to make significant progress in curbing emissions. For example, the government's sustainability advisor, the Sustainable Development Commission, suggested that '. . . current methods of dealing with climate change are incompatible with the task at hand' (SDC 2006a, 6). In response to this criticism, and with growing evidence of the detrimental effects of climate change (IPCC 2007; NCAR 2006), the government introduced a raft of new climate change policies in 2005–2007, including a number of significant initiatives to encourage the development of low carbon homes, such as relief from stamp duty (house sales tax) for zero carbon homes (HM Revenue & Customs 2007), a new national code for the sustainable design and construction of housing (the Code for Sustainable Homes), a pledge for all new housing in England and Wales to be zero carbon by the year 2016, and mandatory energy performance certificates for homes when they are sold (Communities and Local Government 2006; Kelly 2006).

Yet, despite the development of UK climate change policy, there remains a shortage of politically acceptable solutions to the problem, thus leaving the government open to challenges of engaging in policy rhetoric rather than action. In other words, there is a gap between the UK government climate change targets, and the policies in place (RCEP 2000; SDC 2003). Thus there is a strong political drive for solutions: 'ready-made' solutions such as sustainable housing are ideal, and the low carbon discourse coalition has been adept at framing low energy housing innovations in a way designed to appeal to policy makers.

The low carbon housing discourse coalition

The UK low carbon housing discourse coalition comprises a loosely connected group of organisations and individuals united through their use of shared language and ideas about solutions to climate change in the housing sector. There are some inherent difficulties in mapping members of the discourse coalition, because of its fluid membership, but in general terms the low carbon coalition comprises three main types of organisation, those involved in domestic energy efficiency, renewable energy, and construction industry modernisation (see Table 3).

The discursive 'frame' used by the low carbon discourse coalition – its way of describing and seeing the world that helps define its boundaries – is ecological modernisation; 'the discourse that recognises the structural character of the environmental problematique, but none the less assumes that existing political, economic, and social institutions can internalise the care for the environment' (Hajer 1995, 25). A central assertion of an ecologically modern approach is that economic growth is not incompatible with protecting the environment, a 'win–win' scenario (Mol and Spaargaren 2000; Murphy 2000). There is also a strong focus on technology-based solutions to environmental problems (Christoff 2000). As Table 4 demonstrates, the low carbon housing discourse coalition has drawn heavily on these ecologically modern ideas.

The ecologically modern discursive frame has been important to the innovation journey of low energy housing in a number of different ways. First, ecologically modern discourse has had strong appeal to policy makers because, as the examples in Table 4 demonstrate, it has presented low

Table 3. Members of the UK low carbon housing discourse coalition.

Core focus of innovation activities	Examples of discourse coalition member organisations
Domestic energy efficiency	local authorities; the Building Research Establishment (BRE); Housing Associations and The Housing Corporation; the Energy Saving Trust (EST); energy utilities; Energy Efficiency Advice Centres; Department for Environment Food and Rural Affairs (DEFRA); Association for the Conservation of Energy (ACE); Office of Communities and Local Government
Renewable energy	renewable energy companies (e.g. Solar Century); Regional Development Authorities (RDAs); energy utilities; BRE; the Carbon Trust; DEFRA; the Renewable Power Association
Construction industry modernisation	Rethinking Construction (the Housing Forum); Office of Communities and Local Government.; large private sector house builders (e.g. Countryside Properties, Laing); small private sector house builders (e.g. Greenfield Way, Gusto Construction); BRE; English Partnerships (the Millennium Communities Programme); Housing Associations and The Housing Corporation

Table 4. Examples of ecologically modern discourse used by the UK low carbon housing discourse coalition.

'Win–win': no incompatibility between economic growth and environmental protection	Technology-based solutions
'Government is seeking a partnership in the way we "de-carbonise"' our building stock, involving all of the players in delivering greener, better buildings faster' (DTI 2003). Extract from a speech by the Construction Minister, Brian Wilson, at the 2003 Buildings Awards	'We try to promote passive [technologies], so that householders don't even need to know that they're making an environmental saving.' (Interview, sustainability manager in the social housing sector, June 2002)
'the houses are more expensive because they are more popular, because it is a super place to live, well designed very attractive, low energy . . . . There will be a premium on the house prices so the [local] authority will get some of that back through each house that is sold off. *So ultimately it will be self financing*.' (Interview, a local authority manager responsible for planning a large low carbon housing development, December 2002)	'So what we're trying to do on our [housing] developments is, its all in there, you buy the house and its there, you don't have to think about it, you're not even aware of it . . . . And all the technologies are put in in a way that you would be nuts to want to replace them with something else.' (Interview, sustainable housing manager at an environmental charity, June 2002)
'our overriding concern is to . . . encourage those yet to embrace the [sustainability] agenda that sustainable construction can be low or no-cost.' (The Housing Forum 2002, 2)	'[we're] trying to come up with a lifestyle that makes it easier and more convenient to live a lower impact existence, than by using conventional alternatives. So what we're saying is that *if you're prepared to work with the infrastructure we've provided, you can achieve really quite astonishing things*. Its possible to live [at BedZed] and be pretty close to carbon neutrality.' (Bill Dunster, Chief Architect of BedZed housing development, quoted in Lowenstein (2001, emphasis added))

energy housing technologies as capable of mitigating climate change with no negative economic implications. This has helped speed up the innovation journey, encouraging the government to direct money and resources towards low energy housing through a number of government grants and policy initiatives (see for example DTI 2007; Energy Saving Trust 2007; HM Revenue & Customs 2007). Second, and related, the ecologically modern framing of low carbon discourse has been used partly as a way of distancing low energy housing technologies from the social and institutional context in which they were initially developed. The approach and language of the low carbon discourse coalition stands in strong contrast to the deep green values and beliefs of the original 1970s sustainable housing movement. In effect the socially 'alternative' origins of these technologies have been placed outside of the ecologically modern discursive frame, as part of an attempt to reorientate the past trajectory of the low energy housing innovation journey. Thus, for instance, the project developers of the UK BedZed housing development comment how 'BedZed is a long way from the hairshirt living that bedevils green living' (Lowenstein 2001, 16). It is notable though how BedZed's architect also describes the development as having origins in the 1970s socially alternative bioregionalism movement (Lowenstein 2001b); the ecologically modern framing is not without tension and dispute.

The key point here is about the power of discourse to retrospectively change the course of innovation journeys by mapping out different pathways and selectively editing the reasons for, and context of, original innovations. For instance, an ex-local authority energy manager in the

East Midlands, UK, describes shifts in discourse from energy conservation to climate change mitigation as follows:

> 10 or 15 years ago [the local authority energy mangers'] jobs were to save energy, to save money. Over the past few years *they have concentrated on saving carbon*, it still saves energy, it still saves money, *so they just package things differently*. (Interview, December 2002).

Similarly, another local authority energy manager in the East Midlands, who has been working on energy issues within local government for the last 25 years, explains his strategy for accessing funding through using different discourses:

> what happened then was that the government came back from Rio . . . and they said we'll do a green house program in council housing. *So we took our strategy and shoved it through the word processor* . . . . So we developed this technique, as *flavour of the month changes, we took the same strategy and just reordered the priorities*. So I secured about £2 million extra money through competitive bidding. (Interview, August 2002).

People involved in sustainable housing innovation have therefore strategically reframed their activities using discourse as new policy issues have risen up the government's agenda during the course of the innovation journey. These interview extracts also illustrate the dominance of ecologically modern ideas within the low carbon discourse coalition, with no perceived incompatibility between economic growth and environmental protection.

Sustainable housing developments as low carbon storylines

The low carbon discourse coalition has used a number of 'storylines' to convey its ecologically modern message about solutions to climate change in the housing sector, including about smart housing and housing 'lifecycles' (see Lovell 2004). The most relevant storyline to this discussion, because of its clear technology-focus, is about existing sustainable housing developments which have been reframed as 'low carbon' by the discourse coalition. Analysis here concentrates on two especially high-profile developments – the BedZed development in south London (BedZed 2001; BRECSU 2002), and the Hockerton Housing Project in the East Midlands, UK (BRECSU 2000; Vale 2001), which have become metaphors – short-hand discursive techniques – used by the low carbon discourse coalition. It is examined here how these storylines have influenced the innovation journey of UK low energy housing in terms of both technology and policy development.

The sustainable housing developments of BedZed and Hockerton have had extensive coverage in specialist and general media, facilitated by their unusual, eye catching appearance (see Figures 1 and 2). These two developments in particular have acquired almost a celebrity status, with much publicity in mainstream media, not just industry journals (Lovell 2007). The Beddington Zero Energy Development, or 'BedZed', is a housing development in south London; the outcome of a joint initiative between the architect Bill Dunster, the Peabody Trust (a Registered Social Landlord), and the environmental consultancy BioRegional Development Group (BRECSU 2002; Lowenstein 2001a). It comprises 82 homes; nearly half of which have been sold on the private market, and the remainder are social or low-income housing. BedZed comprises a number of sustainable innovations including an on-site combined heat and power plant, an electric car pool, rainwater tanks, and sedum grass roofs (BRECSU 2002). Hockerton is an earth-sheltered housing development near Newark in the East Midlands. The five terraced homes have no need for central heating: large conservatories collect heat from the sun, and the walls are very well-insulated.

Figure 1. The BedZed housing development, London.

Figure 2. Hockerton Housing Project, Hockerton.

Electricity is provided by photovoltaic panels and a wind turbine, and all wastewater is treated on-site in a reed bed (BRECSU 2000). Both Hockerton and BedZed comprise a number of social or administrative innovations, ranging from the community leasehold at Hockerton to the electric car share scheme at BedZed. However, it is their technical energy innovations that have been prioritised by the low carbon discourse coalition, a point returned to below.

BedZed and Hockerton have acted as a focal point for policy makers, uniting otherwise disparate actors, and thereby have created further opportunities for innovation. Table 5 gives examples of UK policy documents citing BedZed. Notably, the policy documents neglect to mention how BedZed emerged in the absence of significant direct government support. The government has nevertheless attempted to associate itself with BedZed and claim ownership through frequent references to the development in policy discourse, and through site visits, including the launch of new policies

Table 5. Examples of UK policy documents citing BedZed.

Policy document	Reference to Bedzed
Speech by Energy Minister Brian Wilson, February 2002 (DTI 2003b)	'Demonstrations such as the developments ... at BedZed ... prove that the technologies are available to deliver practical systems'
Royal Commission on Environmental Pollution, 22nd Report: 'Energy – Our Changing Climate'	Has a case study box devoted to BedZed and describes it as: 'the most ambitious low energy housing development in the UK to date' (RCEP 2000, 105)
Government Energy Efficiency Best Practice Programme – General Information Report no. 89	'... BedZed represent[s] state-of-the-art for sustainable housing in the UK' (BRECSU 2002, 3)
UK 2003: The Official Yearbook of the United Kingdom of Great Britain and Northern Ireland (The Stationery Office, 2003)	Double page picture spread (pp. 298–99)
Environment Agency report – 'Our Urban Future', September 2002 (www.environment-agency.gov.uk)	BedZed is cited as an example of a solution to climate change
The Housing Corporation (2004)	It is used as a model case study for Registered Social Landlords '... to show how sustainable development can be achieved'

there. For example, Patricia Hewitt, the then UK Secretary for Trade and Industry, used BedZed as a location to announce a new government solar power initiative (DTI 2002). Similarly, the Liberal Democrat party leader visited because he 'was making an environment announcement later that day and wanted a photo to go with any publicity' (BioRegional Communications Officer 2004, personal communication).

The storylines about these two housing developments have promoted further low energy housing innovation, and thereby helped speed up the innovation journey, because their material existence has acted to convince others – government, business and householders – of the economic and technical feasibility of low carbon housing. As the project manager of the Hockerton housing project succinctly summarises:

> The most effective tool is the place and the fact that we are living in it. (Interview, Hockerton Project Manager, March 2003)

In other words, the innovative dwellings literally lend substance to the discourse coalition's arguments, and have helped sustainable housing to be viewed as a credible solution to climate change. Similarly, the authors of the Government-commissioned report about BedZed stress how the development represents:

> ... a powerful argument for the feasibility of a zero-carbon target for all new build. (BRECSU 2002, 11)

and a housing policy officer at a non-governmental organisation involved in sustainable housing describes how they have used BedZed to influence key decision makers in government:

> I think [exemplar projects] are invaluable for showing people what might be done. It is really great when we want to talk to people about sustainable housing – important people – we take them to

> BedZed . . . and to actually see it in action I think is very inspiring, rather than just talking about what it might look like. (Interview, Sustainable Homes co-ordinator at a national environmental NGO, May 2003)

These technology-focused storylines thereby constitute a key element of the ecologically modern low carbon discourse used by the coalition. The interview extracts also illustrate how, with the rise of low carbon discourse in the UK housing sector (Lovell 2004; Toke 2000), it has become increasingly important not just to participate in the discourse, but to have actual material evidence of low carbon practice in order to promote new ideas and gain support. A small number of policy theorists have considered the impact of seeing new policy ideas and innovations in practice (see Guy and Osborn 2001; Owens and Rayner 1999). Hajer, for example, discusses the key role of policy actors' excursions to certain sites of interest, in particular when visiting forests in Scandinavia damaged by acid rain:

> A striking finding . . . concerned the role of meetings and excursions in the process of persuasion . . . these practices . . . can . . . be identified as an essential moment in the process of proliferation and utilization of knowledge, and . . . policy change . . .. (Hajer 1995, 271)

Thus actually seeing the damaged trees helped catalyse shifts in policy. Applied to the idea of an innovation journey, it reminds us of the agency of technologies and materials within their own innovation journey. In particular, there is a stage when the physical presence of innovations, and publicity about them, acts to speed up the innovation journey and simultaneously opens up a number of different pathways for future development and growth. Thus there is a two-way process of interaction between innovations and discourse; it is not just that publicity, discourse and the policy context influence the innovation journey, but also that the innovations themselves shape policy and discourse. For instance, both Hockerton and BedZed are cited in policy documents recommending changes to the UK energy building regulations because of climate change (ODPM 2000, 2003a). In addition, new policies have been forthcoming at a local government level, based on the experience of BedZed and Hockerton. Sutton Borough Council, where BedZed is located, set an important new precedent in planning procedure by awarding the development contract to the BedZed team, despite not being the highest bidder (BRECSU 2002). Experience with BedZed also informed the Unitary Development Plan produced by Merton Borough Council – the neighbouring local authority – which now requires new developments over a certain size to source 10% of their energy from renewable resources (Merton Borough Council 2004). The 10% renewable energy requirement, which became known as 'the Merton Rule', was subsequently incorporated into national government planning guidance (Merton Borough Council 2008).

In sum, there have been a number of important implications of the rise in low carbon discourse for the innovation journey of low energy housing. One effect has been to speed up the innovation journey: the low energy housing sector has experienced rapid growth. Surveys have identified over four hundred low energy housing developments or single houses in the UK (White 2002), with nearly 200 built or planned since the late 1990s (Sustainable Homes 2003). This increase in the pace of the innovation journey has partly been achieved through greater publicity of low energy housing innovations, which has encouraged consumer demand (see Lovell 2005b), but also through a number of new government grant programmes for low energy technologies (DTI 2007; see also Energy Saving Trust 2007). A second effect of the spread of low carbon discourse has been to prioritise the development of certain types of low energy innovation over others, notably technical innovations over administrative ones. This relates to the type of ecologically modern discourse adopted by the low carbon discourse coalitions, which has tended to prioritise technical solutions to environmental problems over institutional or social reform (see Table 4). Thus, for example, highly visible renewable energy technologies such as photovoltaic panels and small wind turbines have been prioritised in government funding programmes such as the Low Carbon Buildings Programme, whereas administrative innovations such as planning reforms, or integrating sustainable decision making within private sector house builders – although an active part of the government's agenda – have been relatively overlooked (House of Commons Environmental Audit Committee 2006).

There have also been negative effects on the low energy housing innovation journey arising from how the boundaries of the low carbon discourse coalition have been narrowly defined in ways that ignore issues that contradict its core ecologically modern ideology. The discourse has been uncritically adopted by members of the coalition, including government, which has inhibited learning. The ecologically modern framing, with its focus on profitable low carbon technologies, has resulted in some important behavioural and social issues arising from occupied low energy housing developments being overlooked. For instance, at BedZed problems have occurred with residents using blinds over the windows – in an effort to gain some privacy – but which block sunlight and hence the passive solar heating effect, which is required to maintain comfortable indoor temperatures (Bioregional 2003). Similarly at Hockerton some of the houses have experienced low winter temperatures below 18 degrees centigrade and residents have had to compensate by wearing extra clothing, or in some cases using portable heating devices such as oil radiators and wood stoves (BRECSU 2000). As these examples demonstrate, it is not the case that residents can become low carbon by simply living in these developments – an idea that the ecologically modern discourse promotes – but rather that some degree of active householder participation and involvement is required. Experience from BedZed and Hockerton suggests that these 'softer' social or administrative innovations are a necessary part of their successful functioning. The second main strand of ecologically modern discourse – about the financial benefits of being green, the 'win–win' scenario – has also acted to exclude important contradictory evidence about the difficulty of making low carbon housing financially viable. BedZed, for example, is said to have overrun its budget by £5 million (Clark and Smit 2004; Lovell 2005a), and Hockerton's moderate construction cost was achieved in large part through use of unpaid self-build labour (Energy Saving Trust 2003).

These kind of technical and financial issues – the 'twists and turns' of a typical innovation journey – do not fit well with the ecologically modern discourse used by the low carbon discourse coalition, and hence have not been widely reported (see for example Bioregional 2003; BRECSU 2000, 2002; Minton 2002). Media coverage of BedZed and Hockerton has almost exclusively been positive: 99% of the articles written about BedZed have been complimentary, as have all but two articles or programmes about Hockerton (Lovell 2005a). A lack of honesty about problems with existing low energy housing technologies can lead to insufficient learning, and therefore hamper their innovation journey because there is a risk of the technologies being used in a more widespread manner without the problems being corrected. Rose (1991) has suggested that honest accounts of policies or projects are not necessary for learning, because the main impact of demonstration projects is to generate 'inspirational learning', thereby motivating others to adopt similar approaches; an outcome that can be seen in the case of low energy housing. However, Rose's focus is more on learning about social issues than technical ones: there are perhaps greater limitations to inspirational learning if significant technical problems are replicated.

Conclusions

The case of UK low energy housing illustrates the influence of discourse on an innovation journey, showing, for example, how discourse smooths out the dynamic twists and turns of an innovation journey through retrospectively retelling and simplifying the journey using a particular discursive frame, in this instance ecological modernisation. It has been discussed how discourse may speed up an innovation journey by creating cohesion among actors, and through presenting technologies in a way designed to appeal to policy makers, but the narrow framing of innovation may create problems because key issues and facts are left out. Learning can be inhibited because of a reluctance to engage with the complex, often contradictory dynamics inherent to an innovation journey. Discourse is evidently an important factor in innovation journeys, and more research is needed to explore whether the patterns and processes identified here might be applicable to other cases.

A discourse analysis perspective also raises some critical challenges for notion of an innovation journey. A key methodological challenge arises from considering the implications of the power of discourse to retrospectively remap an innovation journey. It suggests the need for researchers to pay sharp attention to how interviewees might 'tell the story' of an innovation journey – neatening and simplifying it within a particular discursive frame – and thereby missing out important contradictory issues and events. Indeed, research has shown that theories about innovation have been used by practitioners as a type of policy discourse, where innovation has been portrayed in line with certain popular theoretical frameworks, such as transitions theory (Smith and Kern 2007). Methodological rigour is therefore needed to triangulate interviews with other data, and perhaps for political and social scientists to engage with the data-intensive mapping of all possible relevant events in an innovation journey in a manner advocated by Van de Ven et al. (1999). The paper highlights the value of an interdisciplinary approach to innovation journeys: a focus on discourse highlights the critical patterns and dynamics of the broader policy, political and cultural context within which an innovation journey takes place, and hence adds significant insight.

References

Barker, K. 2004. *Review of housing supply – securing our future housing needs. Final Report*. London: HMSO.

Barton, H., ed. 2000. *Sustainable communities: the potential for eco-neighbourhoods*. London: Earthscan.

BedZed, 2001. About BedZed – what is BedZed? http://www.bedzed.org.uk/about.htm.

Bioregional. 2003. *Technical tour of BedZed housing development*. South London: H. LovellSutton.

BioRegional Communications Officer. 2004. Email correspondence: Jennie Organ, Communications Officer – BioRegional Development Group. H. Lovell, London.

BRECSU. 1996. General Information Report 53: Building a sustainable future – homes for an autonomous community. Building Research Establishment (BRE), Watford.

———. 2000. New Practice Profile 119: The Hockerton Housing Project – design lessons for developers and clients. Building Research Establishment (BRE), Watford.

———. 2002. General Information Report 89: BedZED – Beddington Zero Energy Development, Sutton. Building Research Establishment (BRE), Watford.

Bulkeley, H. 2000. Discourse coalitions and the Australian climate change policy network. *Environment and Planning* C, 18: 727–48.

Bulkeley, H., and Betsill, M. 2003. *Cities and climate change: urban sustainability and global environmental governance*. London: Routledge.

CAT. 1995. *Crazy idealists? The CAT story*. Machynlleth: The Centre for Alternative Technology (CAT).

Chappells, H., and Shove, E. 2000. Sustainable homes and integration. In *Domestic consumption utility services and the environment (DOMUS)*, eds. H. Chappells, M. Klintman, A.-L. Linden et al., 105–28. Wageningen: Universities of Lancaster, and Lund.

Christoff, P. 2000. Ecological modernisation, ecological modernities. *Environmental Politics* 5: 476–500.

Clark, P., and J. Smit. 2004. Dream over. *Building* 1–3.

Communities and Local Government. 2006. *Building a greener future: towards zero carbon development – consultation.* London, HMSO.
DEFRA. 2000. Transport 2010 – the ten year plan. Department for the Environment Food and Rural Affairs (DEFRA), London.
Dobson, A. 2000. *Green political thought.* Routledge, London and New York.
Dryzek, J. 1997. *The politics of the Earth: environmental discourses.* Oxford: Oxford University Press.
DTI. 2002. First stage of major PV Demonstration Programme launched. *New Review* 52, no. 1.
———. 2003. *Energy white paper: our energy future – creating a low carbon economy.* London: HMSO.
———. 2007. *Low carbon buildings programme, phase 2.* http://www.lowcarbonbuildingsphase2.org.uk/.
Energy Saving Trust. 2003. *Sustainable construction: Hockerton Housing Project.* www.practicalhelp.org.uk.
———. 2007. *Housing and buildings funding database.* http://www.energysavingtrust.org.uk/housingbuildings/funding/database/.
Findhorn Ecovillage. 2003. Findhorn – an introduction. http://www.ecovillagefindhorn.com/.
Flyvbjerg, B. 1998. *Rationality and power: democracy in practice.* Chicago, IL: University of Chicago Press.
Guy, S., and S. Osborn. 2001. Contesting environmental design: the hybrid green building. In *Infrastructure in transition: urban networks, buildings, plans*, eds. S. Guy, S. Marvin and T. Moss, 87–102. London: Earthscan.
Hajer, M.A. 1995. *The politics of environmental discourse: ecological modernisation and the policy process.* Oxford: Clarendon Press.
Hastings, A. 1999. Analysing power relations in partnerships: is there a role for discourse analysis? *Urban Studies* 36: 91–106.
HM Government. 2006. *Climate change and sustainable energy act 2006.* http://www.opsi.gov.uk/ACTS/acts2006/ukpga_20060019_en.pdf.
HM Revenue & Customs. 2007. *Stamp duty land tax relief for new zero-carbon homes.* http://www.hmrc.gov.uk/ria/9-zero-carbon-homes.pdf.
House of Commons Environmental Audit Committee. 2006. *Sustainable housing: a follow-up report.* London: HMSO, 275.
House of Commons. 2006. *Climate change and sustainable energy bill.* http://www.publications.parliament.uk/pa/cm200506/cmbills/017/2006017.pdf.
IPCC. 2007. Climate *change 2007: the physical science basis – summary for policymakers.* Paris: Intergovernmental Panel on Climate Change (IPCC).
Kelly, R. 2006. *Shaping a low carbon future – our environmental vision.* http://www.communities.gov.uk/index.asp?id=1505202.
Laws, D., and M. Rein. 2003. Reframing practice. In *Deliberative policy analysis: understanding governance in the network society*, eds. M.A. Hajer and H. Wagenaar, 172–206. Cambridge: Cambridge University Press.
Lounsbury, M., M.J. Ventresca, and P.M. Hirsch. 2003. Social movements, field frames and industry emergence: a cultural-political perspective on US recycling. *Socio-Economic Review* 1: 71–104.
Lovell, H. 2004. Framing sustainable housing as a solution to climate change. *Journal of Environmental Policy and Planning* 6: 35–56.
———. 2005a. *The governance of emerging socio-technical systems: the case of low energy housing in the UK.* Department of Geography, Cambridge University, Cambridge.
———. 2005b. Supply and demand for low energy housing in the UK: insights from a science and technology studies approach. *Housing Studies* 20: 815–29.
———. 2007. The governance of innovation in socio-technical systems: the difficulties of strategic niche management in practice. *Science and Public Policy* 34: 35–44.
Lowenstein, O. 2001a. From BedZed to eternity. *Building for a Future* 11: 16–21.
———. 2001b. Zero energy homes come to the United Kingdom. *Building for a Future* 11: 8–14.
Marsh, D., and R.A.W. Rhodes, eds. 1992. *Policy Networks in British Government.* Oxford: Clarendon Press.
Merton Borough Council. 2004. *Forum for the Future – a radical planning decision to promote renewable energy.* http://www.regionalfutures.org.uk/newsdigest/mertonboroughcouncil_page2181.aspx.
———. 2008. The 10% renewable energy policy. http://www.merton.gov.uk/living/planning/plansandprojects/10percentpolicy.htm.
Minton, A. 2002. Eco experiments. *Estates Gazette* 62–3.
Mol, A.P.J., and G. Spaargaren, G. 2000. Ecological modernisation theory in debate: a review. In *Ecological modernisation around the world: perspectives and critical debates*, eds. A.P.J. Mol and D.A. Sonnenfeld, 17–49. London: Frank Cass.
Murphy, J. 2000. Editorial – ecological modernisation. *Geoforum* 31: 1–8.

NCAR (US National Centre for Atmospheric Research). 2006. Abrupt ice retreat could produce ice-free arctic summers by 2040.

Newell, P. 2000. *Climate for change: non-state actors and the global politics of the greenhouse*. Cambridge: Cambridge University Press.

ODPM. 2000. Current thinking on possible future amendments of energy efficiency provisions. http://www.odpm.gov.uk/stellent/groups/odpm_buildreg/documents/page/odpm_breg_600333.hcsp.

———. 2003a. Industry Advisory Group (Working Party 4). Low and Zero Carbon Energy Systems Meeting Minutes, 16 October 2003. www.odpm.gov.uk.

———. 2003b. *Sustainable communities: building for the future*. London: HMSO.

Owens, S., and T. Rayner. 1999. When knowledge matters: the role and influence of the Royal Commission on Environmental Pollution. *Journal of Environmental Policy and Planning* 1: 7–24.

Owens, S.E., and R. Cowell. 2002. *Land and limits: interpreting sustainability in the planning process*. London: Routledge.

Porter, G., and J.W. Brown. 1996. *Global environmental politics*. Boulder, CO: Westview Press.

RCEP. 2000. *Energy – the changing climate, 22nd Report*. London: HMSO.

Rein, M., and D. Schon. 1993. Reframing policy discourse. In *The argumentative turn in policy analysis and planning*, eds. F. Fischer and J. Forester, 145–66. London: UCL Press.

Rose, R. 1991. What is lesson drawing. *Journal of Public Policy* 11: 3–30.

Rydin, Y. 1999. Can we talk ourselves into sustainability? The role of discourse in the environmental policy process. *Environmental Values* 8: 467–84.

Sabatier, P.A. 1999. *Theories of the policy process: theoretical lenses on public policy*. Boulder, CO: Westview Press.

Sabatier, P.A., and Jenkins Smith, H.C., eds. 1993. *Policy change and learning: an advocacy coalition approach*. Boulder, CO: Westview Press.

SDC (Sustainable Development Commission). 2003. Policy audit of United Kingdom climate change policies and programmes. SDC, The Edinburgh Centre for Carbon Management Ltd, London.

———. 2006. Climate change: the UK programme – an SDC response. http://www.sd-commission.org.uk/publications.

Sharp, L., and T. Richardson. 2001. Reflections on Foucauldian discourse analysis in planning and environmental policy research. *Journal of Environmental Policy and Planning* 3: 193–209.

Smith, A. 2003. Transforming technological regimes for sustainable development: a role for alternative technology niches? *Science and Public Policy* 30: 127–35.

———. 2004. Governance lessons from green niches: the case of eco-housing. Paper presented at ESRC Sustainable Technologies Programme Workshop on Governance, Technology and Sustainability, Open University, September 7–8, Milton Keynes.

———. 2007. Translating sustainabilities between green niches and socio-technical regimes. *Technology Analysis & Strategic Management* 19, no. 4: 427–50.

Smith, A., and F. Kern. 2007. The transitions discourse in the ecological modernisation of the Netherlands. SPRU Working Paper no.160, SPRU, University of Sussex, Brighton.

Smith, M., J. Whitelegg, and N.J. Williams. 1998. *Greening the built environment*. London, Earthscan.

Stern, S.N. 2006. *The stern review on the economics of climate change*. London: HM Treasury.

Sustainable Homes. 2003. Sustainable Homes Ecodatabase. http://www.sustainablehomes.co.uk/.

The Housing Forum. 2001. *Homing in on excellence: a commentary on the use of offsite fabrication methods for the United Kingdom house building industry*. The Housing Forum, London.

Toke, D. 2000. Policy network creation: the case of energy efficiency. *Public Administration* 78: 835–54.

Turnpenny, J., A. Haxeltine et al. 2005. Mapping actors involved in climate change policy networks in the UK. Tyndall Centre Working Paper no. 66, January 2005.

Vale, R. 2001. Hockerton Housing Project – Case Study 7.7b. Australian Greenhouse Office, Institute for Sustainable Futures, Canberra, 1–6.

Van de Ven, A.H., D.E. Polley, R. Garud, and S. Venkataraman. 1999. *The innovation journey*. Oxford: Oxford University Press.

Vidal, J. 2006. The year the world woke up. *The Guardian*, December 20.

White, N. 2002. Sustainable housing schemes in the United Kingdom, Hockerton Housing Project, Hockerton.

Wood, A. 1990. History and overview of communal living, http://www.diggersanddreamers.org.uk/Articles/199001.htm.

Young, S.C. 2000. Introduction: the origins and evolving nature of ecological modernisation. In *The emergence of ecological modernisation: integrating the environment and the economy?*, ed. S.C. Young, 1–39. London: Routledge.

Socio-political embedding of onshore wind power in the Netherlands and North Rhine–Westphalia

Susanne Agterbosch and Sylvia Breukers

This paper compares the historical socio-political innovation journeys of onshore wind power in North Rhine–Westphalia and the Netherlands, concentrating on the implementation (realisation) of wind energy projects and the market of project development. Attention is drawn to the level of implementation (where struggles between multiple interests and meanings become manifest) and to the ability of entrepreneurs to succesfully plan and develop projects. A historical new-institutionalist approach is adopted, in which actors, networks and their institutional environment are understood as mutually constitutive. We investigate how socio-political embedding has come about – the process through, which a new technology becomes embedded in existing and changing rules and routines of society. A conclusion is that socio-political embedding is crucial to the implementation achievements and market developments. The legitimacy of wind projects is not self-evident. Addressing socio-political embedding contributes to a better understanding of the development of this legitimacy. Not only researchers but also policy makers should address legitimacy as a central issue, not just an afterthought to the development and diffusion of new technologies. A policy and planning strategy that involves the institutionalisation of early participation of relevant stakeholders can enhance the legitimacy of both the process and outcome and contribute to social-political innovation needed to accomplish sustainable innovation journeys.

1. Introduction

Whether an innovation journey is sustainable or not, depends not just on the technological innovation and its economic–technical performance. It also depends on how this journey is organised, who is invited to travel along, and on how the impacts of the journey on others and on the natural environment are dealt with. Traveling along a novel technological path has an inherent socio-political dimension, and this stands central in this paper. We present a historical comparison of the socio-political innovation journeys of wind power onshore in the Netherlands and the German state of North Rhine–Westphalia (NRW).

Sustainability has ecological, economic and socio-political dimensions. Many studies (e.g. Junginger, Faaij, and Turkenburg 2004) investigate how the first two dimensions can support each other (e.g. aiming at a win–win situation), reducing the third one to the interactions between the main actors that are directly involved in the development and application of new technologies. Against this we argue for a more profound elaboration of the socio-political potential of environmental technologies.

Studies that address innovation journeys from a system's approach increasingly recognise the dynamic and interrelated social and institutional conditions that affect technological innovation (Hekkert et al. 2007; Negro 2007; Geels 2004). As for the diffusion of wind power technology, several systemic studies investigate the interactions between actors, networks and institutions (Bergek and Jacobsson 2003; Hofman 2005; Van Est 1999). These studies vary in the importance they attribute to the legitimacy and social acceptance of wind power. Overall, they concentrate on the diffusion and market development of the wind turbine technology. Such a focus makes it difficult to investigate social acceptance and socio-political embedding, because for that one needs to inquire into the processes aimed at the realisation of wind energy projects. This paper presents such an inquiry: it concentrates on the implementation (realisation) of wind energy projects and the market of project development, and as such takes a different angle than studies focussing on turbine technology diffusion.

The relevance of implementation lies in the fact that ultimately, the outcomes of all processes aimed at implementing wind projects make up the aggregated installed capacity at national or state level. The market of project development is characterised by a differentiation in types of entrepreneurs that invest in wind projects, which has hardly been problematised in innovation studies. This is surprising because understanding conditions that prompt entrepreneurs to invest in projects and understanding the conditions that influence their success in realising their projects in a legitimate way is vital for successful market development (Agterbosch 2006).

1.1. *Socio-political embedding and market development*

Wind power implementation can be understood as part of an effort to institutionalise environmental considerations and integrate them into prevailing processes of electricity production and consumption. However, wind power (like other renewable technologies) has to contend with sometimes deep-rooted perceptions, values and habits – of a variety of actors at different scale levels. At the level of the energy system, wind power implementation constitutes challenges to the material infrastructure, and to the rules that govern behaviour in the incumbent energy system. At the level of implementation, wind turbines have a significant spatial impact, because of their physical appearance, but also because of wind power's decentralised characteristics and its resource dependent nature: a lot of sites are needed and many people will be confronted with the turbines. The most salient concerns in evaluating a wind project on its costs and benefits involve landscape values (Wolsink 2007). Wind projects are susceptible to 'unequal' distribution of costs and benefits, whereby the costs (visual impact) are borne by the local community while the benefits are for the developers and society at large (financial gain and CO_2 reduction) (Wolsink 2000). Negative attitudes towards wind projects are moreover intensified by discontent with the planning, decision-making on and management of wind projects (Pasqualetti, Gipe and Righter 2002).

Problematising socio-political legitimacy of policy- and implementation processes stands central in spatial planning, (environmental) policy sciences and technology assessment studies (Arts and Leroy 2006; Garud and Ahlstrom 1997; Hajer and Wagenaar 2002; Healey 1997; Hill and Hupe 2002). For wind power, an inquiry into the socio-political processes involved in the adoption, implementation and market development can contribute to a better understanding of diverging achievements between geographical cases. Special attention is needed for the operational level of implementation, because there the struggles between multiple interests and meaning become manifest. We intend to draw conclusions on the relationship between socio-political embedding and market development, on the basis of extensive empirical research (Agterbosch 2006; Breukers 2006). Our main questions is therefore:

- *How can we understand the relationship between socio-political embedding and market development?*

2. Theoretical and conceptual framework

Various disciplines like economics, political science, planning theory, sociology and comparative historical analysis have conceived a renewed interest in how the role of institutions can contribute to the understanding of social and political outcomes (Hall and Taylor 1996; Powell and DiMaggio 1991). Although a strict delineation of different new institutionalist approaches is difficult, our approach is most in line with historical new institutionalism (Thelen 1999). Historical institutionalism addresses unintended consequences as well as inefficiencies generated by institutions (and as such it differs from rational-choice approaches that regard institutions as purposive and efficient). Historical institutionalism is interested in how institutions provide or impede access to decision making for different groups in society and how this changes over time (Hall and Taylor 1996). Functionalist, political and cultural explanations can all account for institutions evolving or changing (Powell and DiMaggio 1991; Scharpf 2000). Behaviour is understood as being motivated by both strategic calculation and normative scripts or 'rules of appropriateness'.

In line with this approach, we define institutions as rules, patterns or procedures that structure behaviour and interaction. These rules can be informal (norms, habits and customs) or formal (written laws, regulations and standards). Actors, networks and their institutional environment are understood as mutually constitutive: while institutions shape and constrain the strategies and options of actors, these institutions themselves can be challenged, altered or reproduced by actors in processes of (strategic) interaction and in political conflict. Having briefly presented our theoretical perspective, we can now proceed to elaborate our conceptual framework.

2.1. *Conceptual framework*

We conceptualise the institutional environment and the practices related to implementation of wind energy as follows. The challenges of wind power implementation are bound up with socio-political processes at different levels of government and transcend the boundaries of established policy domains. Policy and implementation processes involve multiple and interdependent (public and private) actors. In their attempt to influence outcomes, they may form networks or coalitions. Outcomes are the result of interactions between actors, networks and the institutional environment. This institutional environment involves the arrangements, rules, norms of the incumbent energy system (which is not fit for and potentially hostile towards wind power), but also arrangements in the domains of spatial planning and environmental policy. Policies and regulation from these policy domains converge on the operational level of implementation and constitute the dynamic context of formal rules in which wind power entrepreneurs and other stakeholders operate. In addition to these formalised rules, informal norms, agreements and behavioural patterns (e.g. local planning traditions; grass-roots environmental activism) may affect wind power implementation. This institutional context structures the interactions among actors in policy making, planning, decision making and project development. Institutional arrangements thus influence how decisions are taken and who is involved in or excluded from these processes. Institutions' role in this 'mobilisation of bias' is that they reflect and maintain the existing distribution of power (Schattschneider 1960). However, institutions as well as power relations may change over time, as a result of (inter)actions internal or external to the sub-domain of wind power. Interaction and cooperation need to be understood in this context of unequal and changing power relations.

2.2. *Operationalisation*

Our interest lies with socio-political processes involved in the adoption, implementation and market development of wind energy onshore. *Socio-political embedding* is defined as the process through which wind power, a new technology, becomes embedded in existing and changing rules and routines of politics, policy and society. This process involves multiple stakeholders at multiple levels and domains of society. We address those aspects of institutionalisation that involve the creation of social, economic and political involvement and/or commitment.

Our empirical analysis aims to qualify this socio-political embedding and to establish how it relates to implementation achievements and market development (market of wind energy project development). To analyse this relationship we present an empirical inquiry into the interconnected historical processes in the relevant policy domains, actual implementation of wind projects, the development of the wind power market, the development of networks and mobilisation processes. It is along these lines that we structure this paper and therefore we first explain briefly how we understand these dimensions.

Policy processes in relevant policy domains and implementation

Policy formation is an interactive political process in which problems and solutions are formulated. Depending on the configuration of actors involved and their interaction (cooperation and/or conflict), a policy issue is framed in a certain manner. Policy outcomes depend on how resources, interests and values are represented in the configuration of the actors involved. Policies provide rules, incentives and constraints that are relevant for those involved in wind power planning and investment decisions, but affect other stakeholders and interests as well. Policies with direct relevance for wind power development come from the domains of *energy policy, spatial planning and environmental policy* at multiple levels of government: EU, national, provincial and municipal policies. Policies are intended to have a particular impact at the level of implementation, but in practice may work out quite differently. Our historical analysis addresses how early policy choices have set in motion a certain path dependence.

Market development and implementation

We address the development of the market in project realisation. Institutional arrangements affect the proliferation of certain types of entrepreneurs over time. On their turn, these entrepreneurs crucially affect how the market evolves, as they invest in and develop projects in different ways. The capacity to implement projects is the *implementation capacity* and differs for different types of entrepreneurs. Implementation of wind power takes place at the local level. At the local level, the final decisions are taken: the decision of the entrepreneur(s) to invest and develop; the decision of the local authorities to grant permission and decisions of local community members to accept or oppose the project. Local planning and decision making around wind projects involves a balancing of diverging interests (local, global, economic, environmental, social, cultural). Not surprisingly, it is at the local level where conflicts of interests most often become manifest.

Network development and mobilisation

For both policy formation and implementation processes we address how processes have proceeded and who has been involved in these. We deal with the formation of *networks* as well as how these have attempted (and succeeded or not) at mobilising support, within the institutional contexts. We can distinguish between socio-political and economic mobilisation, which takes place through cooperation and interactions at several levels. Socio-political support can be mobilised at the local level of implementation but also in lobbies aimed at changing government policies. The same goes for economic mobilisation, e.g. mobilising capital from local investors or by making claims on national financial support budgets. Socio-political and economic mobilisation are intertwined, for instance, the inviting of local financial participation in wind projects may create both economic and socio-political mobilisation.

2.3. *Multiple case method*

Our analysis is based on a multiple embedded case study (Yin 1994) on the implementation of wind power onshore in the Netherlands and the German state of North Rhine–Westphalia (NRW) over the period 1989–2004. We used data from policy documents, stakeholder documents and meetings, and interviews with key stakeholders in all relevant policy domains. Respondents included people from conventional energy-sector branches, wind project developers, environmental, nature protection and landscape preservation organisations, anti-wind power groups, research institutes, and policy makers and planners at different levels of government (Agterbosch 2006; Breukers 2006).

The selected cases differ in terms of achievements in wind power implementation, but are comparable in other aspects. Both cases are subject to EU law and regulation. The national governments of both cases have (at least in words) committed themselves to the development of wind power since the oil crisis in 1973. In addition, in each case the general public attitude towards wind power has overall been positive.

Until around 1999, the Netherlands were ahead of NRW in terms of absolute numbers in installed capacity (Table 1). Developments started earlier, but from early on met with local resistance, especially when larger-scale projects were initiated. As a result, many wind projects have never been realised (Agterbosch, Glasbergen and Vermeulen 2007; Breukers and Wolsink 2007b; Verbong 1999). Developments in NRW started later and project development built up through grass-roots initiatives rather than by attempts at large-scale projects. While throughout the nineties,

Table 1. Cumulative installed capacity (MW).

	1992	1995	1998	2000	2004
Netherlands	57	257	364	442	1,080
NRW	±15	110	326	644	2,053

NRW implementation rates surpassed targets set by the state government, Dutch implementation soured and remained behind targets set by the government. After 1999, implementation in NRW accelerated and surpassed the achievements in the Netherlands.

In terms of the social acceptance of wind energy projects, NRW has a better record than the Netherlands. In NRW, opposition did not hamper implementation from early onwards, but became significant much later – especially after 2001. Overall, implementation achievements in NRW can be characterised as impressive compared with those in the Netherlands. More installed capacity has been realised in a less discordant climate. The relative success of NRW is furthermore affirmed when taking into account that it has less favourable wind conditions, about the same area surface and a slightly higher population density compared with the Netherlands (Breukers 2006).

We chose to NRW as a case instead of the whole of Germany because the different German federal states have very diverse histories in terms of policies, laws and implementation achievements. Still, we address the federal context of policy making and politics, but limit the account of sub-national developments and achievements to those pertaining to NRW.

3. Case studies: the Netherlands and North Rhine–Westphalia

Building on research conducted by Agterbosch (2006) and Breukers (2006), this section presents comparative accounts, which first address developments in the relevant policy domains of energy, spatial planning and environmental policy. Next we turn to market development and implementation, followed by a section on network development and mobilisation.

3.1. *Energy domain and implementation*

Netherlands

In the first half of the 1990s, the Dutch wind power supply market was monopolistic. Private wind power producers could sell their electricity output to the regional energy distributors. The latter were required by law to purchase all the electricity generated by private wind power producers located in the area in which they had a monopoly on supply. The private wind power producers had to negotiate the price they received for wind-generated electricity with energy distributors. Because distributors were also developing wind projects, the private wind producers in fact had to negotiate the price with their direct competitors. In 1992, a fierce discussion about these remunerations erupted between distributors and the Union of Private Wind Turbine Operators (PAWEX). The distributors won so that they could effectively define the remunerations. Because the 1989 Electricity Act allowed them to set different arrangements for different technologies, they could penalize intermittent sources such as wind power.

In the light of the first National Environmental Plan (1989) the government defined energy distributors as a 'target group' with which voluntary agreements were set: the Environmental Action Plans (MAPs). These covenants (1991–2001) allowed distributors to impose a 'MAP levy' on electricity consumption and employ the resulting funds to support renewables such as wind power. The MAPs favoured distributors over private entrepreneurs because the latter had no legal entitlement to MAP funds. With this and other policies, the government effectively was drawing out the course that was to be taken in wind power development, awarding the energy sector a central role.

In the second half of the 1990s, two interrelated changes affected wind power implementation. First, a change from a monopolistic market structure, in which strategic energy policies were devised in close cooperation with the energy sector, to a liberalised market structure, with less interference of the energy sector in renewable energy policy making. The liberalisation process enhanced the bargaining powers of private entrepreneurs, who were no longer obliged to sell their electricity to the regional energy distributors. In addition, their guaranteed and immediate access to the grid was now formally arranged.

A second change was the change towards a financial support system that no longer favoured energy distributors over other entrepreneurs involved in wind power development. Investment subsidies that had been in place so far were considered incompatible with the trend of market liberalisation. From 1996 onwards, these were gradually replaced by fiscal instruments. An attractive investment climate was created as well as an improved level playing field. The improved investment environment encouraged new independent wind power producers to rise in the late 1990s (Agterbosch 2006). In 2003, part of the fiscal support was replaced by the Environment Quality of Production (MEP), a policy that involved a feed-in tariff component for domestic producers which furthermore encouraged private wind power initiatives.

In 2001, a green certificate system replaced the voluntary Green Label system (implemented by energy distributors in 1998). This and the termination of the MAPs in 2000 implicated that energy distributors lost their privileged position in receiving financial support for wind power developments. As a result, their implementation capacity decreased in comparison to the implementation capacity for private power producers.

North Rhine–Westphalia

The adoption of the feed-in tariff system in 1991 and its continuation until today has been vital for the developments in NRW. However, even before the feed-in tariff system was introduced, NRW adopted a progressive integrated energy program in 1987 that supported small-scale initiatives in turbine and project development. The decentralised administrative structure enabled state governments to adopt innovative policies that aligned well with practices within their territories. The NRW approach aimed both at supporting the ailing coal industry (for employment reasons) while at the same time encouraging new and clean energy technologies (to trigger the rise of a new technological sector that would also create employment). Hence, the state government from early on attempted to combine environmental, economic and employment concerns.

The feed-in tariff system obliged the utilities to provide access to the grid and to pay a feed-in price on top of the normal electricity price. There was no limit to the amount of renewable electricity that could be fed into the grid. Because initially the law did not apply to renewable electricity from applications that were for 25% or more publicly owned, the utilities would not receive extra benefits when producing renewable electricity themselves.

Federal and NRW level policies and the guaranteed grid access stimulated diversity and supported grass-roots initiatives. This triggered technological innovation, market development and an early increase in installed capacity, mainly by farmers and through locally owned projects. Private capital was mobilised on a large scale, from farmers, small companies and individuals. A firm base was created for both implementation of wind power and a flourishing home market for turbine manufacturers and related services and industries (Breukers 2006; Jacobsson and Lauber 2006).

In the course of the 1990s, the configuration of types of entrepreneurs active in wind project development changed because of the professionalising market. Citizens' wind projects made place for projects developed by medium-sized companies – sometimes selling shares to citizens, but not necessarily. The implementation capacity of the latter increased while citizens' projects diminished in importance.

3.2. *Domain of spatial planning and implementation*

Both the Dutch and German planning system can be characterised as decentralised, in the sense that the final decision making on local land use is the competence of the local authority. A pro-active decision of the local authority in principle is needed before any permitting procedure can start, because a change in the local land use plan is required that allows for wind power developments.

Netherlands

Local planning and securing sites and permits for wind projects was problematic from early on. In 1991, the ministries of Economic Affairs and Housing, Spatial Planning and the Environment and seven coastal provinces signed an administrative covenant in order to improve the siting of turbines. The covenant contained a national target of 1000 MW to be installed by 2000, with specified targets for each province. Municipalities were not involved in the covenant, neither were other stakeholders, e.g. project developers, environmental and nature protection organisations. These targets were not reached. Central and provincial authorities focused on large-scale applications, to be developed by energy distributors. However, the distributors lacked experience in planning of decentralised facilities and often adopted a project planning approach that ignored local level concerns and discussions (e.g. landscape impact and local economic benefit). This created conflicts at the level of local politics (Agterbosch 2006; Breukers and Wolsink 2007b).

When a Dutch municipality does not take a pro-active decision to adapt the local land use plan, a wind power entrepreneur is left empty-handed. This is what often has happened under the influence of local politics. Although the permission procedures are complex, most projects fail during the local political debate prior to the actual permission procedures (Agterbosch, Glasbergen, and Vermeulen 2007).

Throughout the 1990s, the Ministry of Economic Affairs urged the designation of sites by provinces or central government, to overcome local opposition against wind projects. However, the Ministry of Housing, Spatial Planning and the Environment – competent to decide on such matters – as well as the provinces opposed such a coercive approach, as it runs counter to the Dutch tradition of covenants and consensus seeking. Instead the 1991 covenant was renewed in 2001. The provinces were granted a more prominent role in encouraging the proliferation of wind projects.

Compared with the distributors, farmer–entrepreneurs in the late 1990s were more successful in overcoming siting problems because they were landowners and better acquainted with local social networks. Their projects represented a local economic benefit. Implementation by these actors accelerated, especially after 2001.

North Rhine–Westphalia

As in the Netherlands, German spatial planning does not provide for participation beyond formal consultation on ready-made plans. There is hardly a tradition of public involvement in local planning. Financial or planning participation in wind projects has been the result of the approach adopted by project developers, not of planning institutions (Breukers 2006).

The permission procedures in NRW overall are not regarded as very complex. Since 1996, NRW Ordinances stipulate unambiguously the criteria that must be met. Unlike the Dutch municipalities, their German counterparts have been required by law to designate locations for wind power developments in their local land use plans since 1997 – the *Priviligierung* or privileging of wind turbines. Some municipalities consider this requirement to take a pro-active stance as an unacceptable infringement on the traditional rights of local self-government in Germany. On the other hand, it can be argued that it is still the municipality (and not a higher tier of government) that decides on the designation. In doing so, it can effectively exclude areas, by setting stringent criteria that must be met by the developers.

The 1997 *Priviligierung* accelerated implementation but also triggered resistance from municipalities. In addition, a changed feed-in tariff system in 2001 made investments in wind projects at sites with moderate wind conditions more attractive, and in NRW this resulted in an unprecedented 'rush' in wind project developments. Fewer projects were developed with local financial participation; the decreased local involvement and the confrontation with the acceleration in project development led to rising concerns and opposition from local citizens, planners, nature protection organisations and municipalities. The privileges attributed to wind turbines were increasingly regarded as going against the interests of landscape and nature protection (Breukers 2006).

3.3. *Environmental policy domain*

Netherlands

The adoption of environmental objectives in policy has had both direct and indirect effects on wind power developments in the Netherlands. Responsibility for wind energy policy was vested with the Dutch Ministry of Economic Affairs. The Ministry of Housing, Spatial Planning and the Environment has not been pro-active in policy making for wind power. The 1989 National Environmental Policy Plan gave environmental concerns more prominence and introduced the new strategy of covenants with industrial target groups. These MAPs however have merely solidified the position of the distributors *vis-à-vis* independent wind power entrepreneurs, although the latter were generally more environmentally motivated and committed to wind power development.

Around 1997, calls for institutionalising a 'Green Polder Model', next to the social-economic polder model were heard, partly in reaction to plans for large-scale infrastructure projects that invoked protests from several stakeholders. The Dutch green party (Green Left) was at its height, but overall its political influence remained very limited. The Green Polder Model was never institutionalised and interest in environmental issues waned around the turn of the century.

The Fourth Environmental Policy Plan (2001) launched a new policy concept of *transition management*, involving a long-term approach to accomplish system innovation – a fundamental change in production chains. The Energy Report (2002) adopted the term energy transition management, referring to a governance approach that invites stakeholder participation and emphasising the idea of co-optimising economy and environment. However, for onshore wind power the Ministry of Economic Affairs has continued to argue in favour of overruling municipalities and limiting formal participation in order to facilitate the siting of turbines. Such an approach has proven to work out negatively for wind power implementation, but also casts doubts on the credibility of claims to a more collaborative policy style (Breukers and Wolsink 2007a).

North Rhine–Westphalia

German environmental policy became entrenched earlier than in the Netherlands. The policy concept of ecological modernisation (decoupling economic growth and pollution flows) was adopted in the early 1980s and new green industries and technologies were supported. Support for wind power fitted well with this, as both economic and ecological aims were pursued.

The early establishment of the Green Party and their entry in the Bundestag in 1983 certainly helped to raise the profile of wind energy. Moreover, because of the federal structure, the Greens could influence policy making by forming coalitions with the social democrat SPD at state level.

In NRW (a traditional SPD-led state) the Greens joined government after 1995. When the Greens became part of the Federal government coalition in 1998, environmental concerns gained more policy attention.

In 2002 the competence for federal renewable policy shifted from the economic to the environmental ministry, which has been positive for the continuation of the feed-in tariff system in the face of unremitting attempts to get rid of this supportive system by both the economics ministry and the conventional energy sector (Breukers 2006; Jacobsson and Lauber 2006).

3.4. *Market development and implementation*

Netherlands

The emergence of the wind power project market in the Netherlands has been the work of four different types of investors: energy distribution companies, small private investors (farmers), new independent wind power producers and wind cooperatives. Each of these actors has been active since the end of the 1980s but they have performed very differently throughout the years. Energy distributors dominated the market in the early 1990s, but their role has declined in the course of the years. Since the late 1990s, farmers have caught up. In 2002 they even surpassed the distributors in terms of total installed capacity. Independent wind power producers also surpassed energy distributors in installed capacity levels over the last couple of years. Only wind cooperatives have remained less important in terms of capacity installed (Agterbosch 2006).

The tendency to facilitate and favour the energy distributors in the early 1990s largely ignored their limited motivation to invest in wind power. In general, distributors did not regard wind power as a technical and economic attractive investment with future potential. The small private investors met with fewer problems at the operational level of implementation compared to distributors, which has added to their success in the late 1990s (Agterbosch, Glasbergen, and Vermeulen 2007; Breukers and Wolsink 2007a).

North Rhine–Westphalia

In NRW the distinction between types of entrepreneurs is less clear cut than in the Netherlands. The most distinctive feature has been the overall absence of the energy companies in the range of entrepreneurs in NRW.

The early market in NRW was characterised by diverse, locally owned projects in which farmers played a major role. The model of *Bürgerwindparks* has been widespread. These wind projects were set up as a company and realised through close cooperation between the initiators (e.g. a company, or a local landowner) and the local community. After 1995, the number of farmer projects decreased, while increasingly small companies (1–3 persons) became involved in planning, developing, and arranging finance for wind projects. They developed turnkey projects or set up a wind project as an independent company in which private individuals and companies could buy shares. Farmers increasingly participated in such initiatives.

By the end of the 1990s, smaller private entrepreneurs decreased in number and those that survived where medium-sized and larger companies. As a general trend, NRW wind projects are increasingly planned without local financial involvement and with less local involvement in project planning.

Various larger wind farms have been developed through investment funds, in which individuals could buy shares while benefiting from an investment tax rebate. Buying shares in wind power funds became very popular around 2000, but in the following years some funds went bankrupt.

This and the insecurity surrounding the renewal of the feed-in tariff system dampened enthusiasm to invest in 2003. Entrepreneurs are increasingly looking abroad to develop projects. In NRW, the best sites – with the least chances of conflict – have already been developed (Breukers 2006).

3.5. *Network development and mobilisation*

Dutch networks involved in mobilising support

The Dutch networks active in wind power development have been fragmented and unsuccessful in mobilising support at the local and national level. The first wind energy pioneers established the Organisation for Renewable Energy (ODE) in the 1970s. Twenty-four wind cooperatives (established in the period 1986–1992) were represented by ODE. They promoted and developed decentralised and small-scale wind power applications. This network did not become the vanguard of Dutch wind power development, because the Ministry of Economic Affairs created its own network around the established energy sector, related agencies and research institutes.

Two different branch organisations were created at the end of the 1980s, one for the turbine manufacturers and one broader association for all kinds of professionals involved in wind power development (NEWIN). Difficulties between energy distributors and independent wind power producers prompted the latter to set up their own organisation in 1992 (PAWEX), explicitly excluding distributors from membership. The three branches had different constituencies and served different and competing interests, which inhibited a joint proliferation and lobby. Mobilising support for wind power at the level of national policy making was unsuccessful. In addition, many members of NEWIN (energy distributors) have also failed in mobilising support at the level of implementation.

At the regional level, four associations of private wind turbine owners were formed at the beginning of the 1990s. Their members (commercially motivated small private investors) pooled resources in regional lobbies, exchanged information and experiences. Because of their capacity to cooperate and to organise themselves, they increasingly managed to strengthen their position in the liberalising market. Many members were farmers and relatively successful in mobilising local support for their project.

Support, scepticism and opposition in the Netherlands

Whereas environmental organisations such as Greenpeace and Friends of the Earth have univocally supported wind power (hardly addressing potential environmental and landscape impacts), organisations such as the Bird Protection Association, the Association for the Protection of the Wadden Sea and the nature protection organisation Foundation for Nature and Environment (SNM) count as conditional supporters. They support wind projects, as long as these meet their criteria regarding landscape, nature and bird protection. The twelve Provincial Environmental Federations – SNM branches – have more reservations. Together with SNM they have identified locations that are acceptable from a nature and landscape protection point of view.

Local opposition against wind projects dates back to the 1980s, when the energy companies took on the development of their first wind farm (which failed). For a long time organised opposition against wind power projects was concentrated in the two provinces of Groningen and Friesland. A National Critical Platform Wind Power was established in 2001, representing more than 40 local anti-wind energy groups in the whole of the Netherlands. Several Dutch academics regularly publish articles that argue against wind power technology in the media. Nature and landscape protection organisations keep their distance from these anti-wind groups and do not wish to be associated with them.

North Rhine–Westphalia: mobilising support through grass roots

In Germany, like in the Netherlands, anti-nuclear and environmental movements inspired people to search for alternatives. The NRW (and German) network around wind power was built up from the bottom upwards. In the 1970s, several organisations were established that represented people active in wind power and other renewables. The German Wind Energy Association was set up in 1985, and has played a decisive role in the development of ideas for policy support, as well as in lobbies at regional, state and federal level.

Several organisations representing interests in renewables were able to mobilise support for renewables among members of the federal parliament, which turned out to be crucial for the adoption and continued support for the feed-in tariff system. Both the energy utilities and the economics ministry have attacked this system throughout the years, politically and in court, but without much success.

Until halfway through the 1990s, many project developers were locally based, e.g. private individuals, farmers and local utilities. Local support was mobilised through the development of citizens' wind projects.

In NRW, companies in component supply as well as the metal industry prospered thanks to the growth of the wind power market. Industry-related actors, ranging from equipment producers to service providers supported wind power lobbies in times of political turmoil (Breukers 2006).

Support, scepticism and opposition in North Rhine–Westphalia

The two nature organisations Nabu and BUND have joined in the support for wind power, but they do not cooperate intensively with environmental organisations such as Greenpeace and the Worldwide Fund for Nature. In recent years, Nabu has become more critical on the accelerated implementation of wind power in NRW, which is perceived as going at the cost of landscape and nature protection values. Moreover, like in the Netherlands, local or regional branches of nature protection organisations in NRW are more sceptical towards wind power development than their national branch.

Citizens' initiatives against wind projects have mushroomed since the late 1990s. After 2001, these have become increasingly adamant in NRW. At the national level, the Darmstadt Manifesto (1998) had been signed by over a 100 academics who requested an immediate withdrawal of financial support for wind power. In addition, some 300 citizens have signed a Resolution of Citizens' Initiatives against Wind Turbines, addressed to all levels of government.

The feeling that in NRW, a certain saturation point has been reached in terms of the number of wind turbines that a landscape can have, has become more prevalent among planners, municipalities and nature protection organisations – actors that are in principle not against wind power. The crucial difference with the Dutch situation is that resistance has become significant at a moment when an impressive installed capacity level had already been reached.

4. Comparison

Energy system

While the German feed-in tariff system was stable over time and focused on yield, the Dutch policy support system was volatile and focused on capital investments without incentives to improve performance. Most policy interventions in the Netherlands were attempts to correct deficient policies or efforts to bring policy in line with the demands of liberalisation. The volatility of the system caused insecurity and negatively affected implementation capacities, in particular for the private and smaller entrepreneurs. For long, Dutch policy awarded a main role in wind project development to the energy sector, which subsequently was not successful in realising wind projects. The institutional framework impeded diversity to occur because a level playing field was absent. This situation was improved when in the second half of the 1990s, policies were introduced that no longer discriminated against private entrepreneurs. The result was acceleration in project implementation. In Germany and NRW, stable policies aimed at diversity and guaranteed grid access and effectively withheld the energy sector from taking on a dominant role. The result was a proliferation of local private initiatives, bringing along local private capital.

Spatial planning

In Germany a pro-active stance of the municipality became required by law (the 1997 *Priviligierung*). In the Netherlands, no such requirement exists. The economics ministry repeatedly proposed measures to overrule local decision making, without success, however. In both cases local level planning and decision making has been characterised by limiting participation to formal consultation on ready-made plans. This means that often, there is no room for deliberation about what are the most relevant decisions of a wind project plan: the location and the size of a project and how this impacts on landscape values and local economic benefits. Where projects have been developed in a participatory manner, this has been the result of the approach of the developers (e.g. the early locally based project developers in NRW), not of planning institutions: it is largely up to the project developer to decide to involve local stakeholders or not.

Environmental policy

Overall, environmental concerns have become institutionalised in society and politics earlier and stronger in NRW and Germany than in the Netherlands. The federal parliament and the environmental ministry have successfully pressured the economics ministry to adopt and stick with the feed-in tariff system. In the Netherlands, no such pressure has been mounted by either the environmental ministry or by parliament. The Dutch progressive environmental policies at the beginning of the 1990s merely strengthened the position of distributors which did not work out positively for wind power implementation.

Market development and implementation

The most important difference between NRW and the Netherlands in entrepreneurship patterns has been the absence of the energy companies as investors in NRW. While in the Netherlands, the dominance of this group impeded the implementation capacities of other entrepreneurs, in NRW a diversity of entrepreneurs was encouraged from early onwards.

The stagnating market in the Netherlands was caused by the inability of the distributors to realise projects, their top–down approach in project development created conflicts at the level of local politics. In the latter part of the 1990s, when investment security was enhanced for a larger diversity of entrepreneurs, smaller locally based private investors rose in importance. This resulted in a significant improvement in the Dutch implementation record.

The role of farmers as entrepreneurs has been important in both NRW and the Netherlands. These farmers own the land on which the projects are built, their projects involve an inherent local benefit, and they are better at home in the local socio-political context and hence often better able to mobilise support for their project. In both the Netherlands and NRW, locally owned projects have proven to be better able to gain local social acceptance. In the Netherlands, this is evidenced by the success of farmers as project initiators in the late 1990s.

Network development and mobilisation

Although both the Netherlands and NRW have seen a grass-roots commitment to wind power development, only in NRW have the early grass-roots initiatives formed the basis for later developments. In the Netherlands their potential remained largely untapped.

The socio-political as well as economic mobilisation has been more successful in NRW than in the Netherlands. This success relates to the fact that support was mobilised through grass-roots initiatives that also were able to jointly gain enough political clout to affect the choices made in policies at NRW and federal level. Economic mobilisation resulted from policies that invited large amounts of private capital to be invested in wind energy. Policies strengthened the implementation capacity of these initially locally based entrepreneurs, and over time this group matured. A flourishing market has come up. The increasing relevance of wind power for employment and industry has been crucial for the broadening of support across political allegiances. At the national level, helped by a diverse lobby and by the Greens at various levels in politics, this embedding was also successful. Throughout the years, the conventional energy companies were never part of the network.

In both the Netherlands and Germany, a neo-corporatist style involved a strong interrelatedness between the economics ministry and the incumbent energy sector. Only in the Netherlands did this relationship determine socio-political mobilisation at the national level in such a way that resulting discriminatory policies in the early years undermined the development of the existing grass-roots basis and contributed to the fragmented character of the representation of wind power interests. Political mobilisation was ineffective, as the wind sector never developed a countervailing power able to influence national policy. The different actors involved in wind power were unable to set aside differences and cooperate: the Dutch networks around wind power failed in mobilising support. Exceptions are some regional associations of small local commercial entrepreneurs – mainly farmers – who did manage to pool resources and improve the mobilisation of support at the local and provincial levels. Economic mobilisation for long remained limited to investments by distributors. Their lack of success in implementation eroded the basis for fostering a domestic turbine manufacturing market. The mobilisation of support *against* wind projects – from municipalities, citizens and increasingly from organised groups – rose much earlier in the Netherlands than in NRW and has contributed to lagging implementation.

5. Discussion: socio-political innovation

Returning to our main question, what can be said about the relationship between socio-political embedding and market development on the basis of our comparative analysis? A mismatch between socio-political processes at the national level and local level, resulting in a partial and insufficient mobilisation process, negatively affects the implementation capacity of entrepreneurs. A lack of socio-political embedding, reflected by rising resistance (at the local level) and half-hearted policies (at the national level), for example, negatively affects implementation and therewith market development.

Our analysis shows that it is useful to relate systemic dimensions to processes at the level of implementation. We have addressed the 'innovation journey' of wind energy from a perspective that focuses on implementation as a socio-political process that involves struggles over interests (e.g. local *vs* national or global interests) and meaning (e.g. turbines as either landscape pollution or symbols of environmental and/or economic progress). The choices and decisions taken at various scale levels are shaped through the interplay between (and exclusion of) several actors, within institutional contexts that change over time (Toke, Breukers, and Wolsink 2008). Obviously, there are different motivations to both support and oppose wind energy projects, and these need not always be clear and transparent. The legitimacy of wind projects is not self-evident. Addressing socio-political embedding – the process through which a new technology becomes embedded in existing and changing rules and routines of society – contributes to a better understanding of the development of this legitimacy and social acceptance. Both researchers and policy makers should address legitimacy as a central issue, not just an afterthought to the development and diffusion of new technologies. A reflexive policy and planning strategy involves the institutionalisation of collaborative learning by involving knowledge, interests and ideas from different stakeholders (e.g. for *ex-ante* evaluation of renewable energy policy). It can enhance the legitimacy of both process and outcomes and contributes to socio-political innovation needed to accomplish sustainable innovation journeys.

References

Agterbosch, S. 2006. Empowering wind power. On social and institutional conditions affecting the performance of entrepreneurs in the wind power supply market in the Netherlands. PhD thesis, Utrecht University.

Agterbosch, S., P. Glasbergen, and W. Vermeulen. 2007. Social barriers in wind power implementation in The Netherlands: perceptions of wind power entrepreneurs and local civil servants of institutional and social conditions in realizing wind power projects, *Renewable and Sustainable Energy Reviews* 11: 1025–55.

Arts, B., and P. Leroy, eds. 2006. *Institutional dynamics in environmental governance*. Dordrecht: Springer.

Bergek, A., and S. Jacobsson. 2003. The emergence of a growth industry: a comparative analysis of the German, Dutch and Swedish wind turbine industries. In *Change, transformation and development: Schumpeterian perspectives*, eds. S. Metcalfe and U. Canter, 197–228. Heidelberg, Physica/Springer.

Breukers, S. 2006. Changing institutional landscapes for wind power implementation. A geographical comparison of institutional capacity building: The Netherlands, England and North Rhine–Westphalia. PhD thesis, University of Amsterdam.

Breukers, S., and M. Wolsink. 2007a. Wind energy policies in the Netherlands: institutional capacity-building for ecological modernisation. *Environmental Politics* 16, no. 1: 92–112.

———. 2007b. Wind power in changing institutional landscapes: an international comparison. *Energy Policy* 35: 2737–50.

Garud, R., and D. Ahlstrom. 1997. Technology assessment: a socio-cognitive perspective. *Journal of Engineering and Technology Management* 14: 25–48.

Geels, F. 2004. From sectoral systems of innovation to socio-technical systems. Insights about dynamics and change from sociology and institutional theory. *Research Policy* 33: 897–920.

Hajer, M., and H. Wagenaar. 2002. *Deliberative policy analysis. Understanding governance in the network society.* Cambridge: Polity Press.

Hall, P.A., and R. Taylor. 1996. Political science and the three new institutionalisms. *Political Studies* 4, no. 4: 936–57.

Healey, P. 1997. *Collaborative planning: shaping places in fragmented societies*. London: MacMillan Press.

Hekkert, M.P., R. Suurs, S. Negro, S. Kuhlmann, and R. Smits. 2007. Functions of innovation systems: a new approach for analysing technological change. *Technological Forecasting and Social Change* 74: 413–32.

Hill, M., and P. Hupe. 2002. *Implementing public policy*. London, Sage.

Hofman, P. 2005. Innovation and institutional change. The transition to a sustainable electricity system. PhD thesis, University of Twente, Enschede.

Jacobsson, S., and V. Lauber. 2006. The politics and policy of energy system transformation. Explaining the German diffusion of renewable energy technology. *Energy Policy* 34: 256–76.

Junginger, M., A. Faaij, and W. Turkenburg. 2004. Global experience curves for wind farms. *Energy Policy* 33: 133–50.

Negro, S. 2007. Dynamics of technological innovation systems. PhD thesis, Utrecht University.

Pasqualetti, M.J., P. Gipe, and R.W. Righter, eds. 2002. *Wind power in view*. San Diego: Academic Press.

Powell, W., and P. DiMaggio. 1991. *The new institutionalism in organizational analysis*. Chicago: University of Chicago Press.
Scharpf, F. 2000. Institutions in comparative policy research. *Comparative Political Studies* 33: 762–90
Schattschneider, J. 1960. *The semi-sovereign people; a realist's view of democracy in America*. New York: Holt, Rinehart and Winston.
Thelen, K. 1999. Historical institutionalism in comparative politics. *Annual Review of Political Science* 2: 369–404.
Toke, D., S. Breukers, and M. Wolsink. 2008. Wind power deployment outcomes: how can we account for the differences? *Renewable and Sustainable Energy Reviews* 12: doi:10.1016/j.rser.2006.10.021.
Van Est, Q.C. 1999. *Winds of change: a comparative study of the politics of wind energy innovation in California and Denmark*. Utrecht: International Books.
Verbong, G. 1999. Wind power in the Netherlands, 1970–1995. *Centaurus* 41, no. 3: 137–60.
Wolsink, M. 2000. Wind power and the NIMBY-myth: institutional capacity and the limited significance of public support. *Renewable Energy* 21: 49–64.
———. 2007. Planning of renewables schemes. Deliberative and fair decision-making on landscape issues instead of reproachful accusations of non-cooperation. *Energy Policy* 35: 2692–04.
Yin, R.K. 1994. *Case study research: design and methods*, 2nd edn. London: Sage.

Index

Page numbers followed by n represent endnotes.
Page numbers in *Italics* represent tables.
Page numbers in **Bold** represent figures.